Regeneration of
Woody Legumes in Sahel

Knud Tybirk

AAU REPORTS 27

Botanical Institute Aarhus University 1991
this issue in collaboration with
Danida Forest Seed Centre, Humlebæk

Contents

Abstract iii
Résumé française iv
Acknowledgements v
Author vi

1. INTRODUCTION 1

2. SEED DISPERSAL 3

    Dispersal syndromes in Sahelian woody legumes 4
        Wind dispersal 4
        Animal dispersal 7
        Water dispersal 9
    Possibilities for dispersal 10

3. SEED PREDATION 13

    Insects on legume seeds 13
        Ecology and biology of bruchids 14
        Lifecycle of bruchids 14
        Host specificity 15
    Predators of Sahelian woody legumes 16
        Bruchid seed predators 16
        Non-bruchid seed predators 16
        Bruchid parasites or predators 17
    Extent of seed predation 17
    Ecological impact of seed predation 21
        Effects of dispersal 21
        Effects of fire 25
        Effects of germination 26

4. GERMINATION OF HARD-SEEDED LEGUMES 29

    Seed characteristics 29
    Seed dormancy 29
    Soil seed bank 32
    Germination in laboratory 33
        Breaking of dormancy 33
        Water imbibition 34
        Influence of pretreatments in laboratory 34
        Influence of other factors in nursery 35

Germination in nature 37
  Effects of dispersal 37
  Effects of predation 38
  Effects of fires 39
  Effects of soil and water 41

5. SEEDLING GROWTH 43

Development of the seedling 43
Vegetative regeneration 44
Factors influencing growth 44
  Water 44
  Soil type 46
  Predation of seedlings 47
  Fire 47
  Competition 48

6. IMPLICATIONS FOR MANAGEMENT 51

7. SPECIES SUMMARIES 53

8. LITERATURE CITED 69

9. GLOSSARY 79

10. INDEX TO SCIENTIFIC NAMES 81

Abstract

This book describes four main aspects of regeneration of woody legumes found in Sahel. The main dispersal strategies are clarified and described and the species are grouped as wind, animal and/or water dispersed. About 50% of the species are primarily wind dispersed, many are dispersed by passage through ungulates, while a few species may also be water dispersed.

The adaptations for dispersal are reflected in the susceptibility for seed predation primarily by beetles of the family Bruchidae. Infestation in freshly collected as well as stored seeds varies from 0 to 100% depending on species and country. Lists of host/predator relationships, levels of attack, *etc.* are included, and ecological aspects of seed predation are discussed in relation to bruchid parasites, dispersal, fire and germination.

Seed dormancy, soil seed bank and different natural and artificial means to overcome the hard seed coat are described. The different treatments applied to different seed lots reflect either different dispersal strategies of the species or different age of seeds. Effects of dispersal, predation, fire, soil, and water on germination are discussed and complex interactions between the ungulate fauna, the grass layer, frequent fires, and bruchid seed predation are discussed in relation to savanna management.

Growth of seedlings as well as vegetative regeneration are described and the main constraints for survival of young plants are fire, drought and browsing. Protection is therefore needed in order to reach higher seedling survival. A list of information on natural and artificial regeneration for each species is included.

It is concluded that any planning of reforestation or change in the management strategies in Sahel should take the natural regeneration and resilience of the vegetation into consideration for long-term planning and sustainable use of the vulnerable areas in Sahel.

Knud Tybirk

Résumé française

L'étude décrit quatre facteurs importants pour la régénération des légumineuses ligneuses du Sahel: les voies de dissémination des graines; les prédateurs des graines; la germination et la croissances des semis.

Les principales strategies de dissémination sont identifiées et décrites et les essences groupées en trois voies principales de dissémination: dissémination par le vent dissémination anthropozoïque et/ou dissémination par l'eau. Environ 50 % des essences sont d'abord disséminées par le vent; beaucoup sont disséminées par l'intermédiaire d'un passage dans les organes digestifs des ongulés; peu d'essences sont aussi disséminées par l'eau.

Les voies de dissémination sont adaptées à l'exposition des graines aux prédateurs, principalement des coléoptères de la famille des Bruchidae. Le taux d'infestation d'une partie des graines nouvellement recoltées ou au dépot varie de zéro à 100 % selon les essences et les pays de récolte. Une liste des prédateurs de graines avec les niveaux d'attaque est élaborée et ventilée selon les essences. Les aspects écologiques des infestations de graines sont discutés pour les bruches en relation avec la dissémination, les feux de brousse et la germination.

La dormance des graines, la banque de graines dans le sol et différents moyens naturels et artificiels pour lever la dormance tégumentaire sont décrits. Les traitements à appliquer pour différents lots de graines sont une conséquence soit de la stratégie de dissémination de l'essence soit de l'âge des graines. Les effets sur la germination de la stratégie de dissémination, de l'infestation, des feux de brousse, du sol et de l'eau sont traités. Les interactions complexes entre les ongulées, la couverture des graminées, les feux de brousse et l'infestation des graines par les bruches sont discutées dans une optique d'aménagement des savanes.

La croissance des semis et la régénération végétative sont décrites. Feux de brousse, sècheresse et broutage sont les principales contraintes pour la survie des jeunes plants. Pour atteindre un meilleur taux de survie, la protection est nécessaire.

Chaque essence est présentée avec une fiche contenant les informations principales concernant la régénération naturelle et artificielle.

On peut conclure que les mécanismes de résilience de la végétation et les facultés de régénération naturelle des légumineuses ligneuses doivent être pris en considération dans toutes les stratégies de reboisement et d'aménagement à long terme dans les zones vulnérables du Sahel.

Acknowledgements

The author wishes to thank Institut des Sciences de l'Environnement and the Departement de Biologie Végétale, Faculté des Sciences, Université C.A.D. de Dakar, Institut Senegalais de Recherches Agricoles, Direction des Recherches sur les Productions Forestieres in Dakar, Centre de Recherches de Zootechnique in Dahra, and Centre de Suivi Ecologique in Dakar for cooperation during the fieldwork for this study.

P. Danthu (ISRA, Direction de Recherches sur les Productions Forestieres, Dakar), S. Langaas (Centre de Suivi Ecologique, Dakar), J. E. Decelle (Musee Royale de l'Afrique Centrale, Tervuren), M. Gissel (Institute of Zoology, Aarhus University), and T. Hauser (Botanical Institute, Aarhus University) are thanked for unpublished results and personal communication used in this book.

Danida Forest Seed Centre is thanked for valuable interest and support of this study. Ivan Nielsen, Jonas Lawesson, and Henrik Balslev Botanical Institute, James L. Luteyn, New York Botanical Garden, Kirsten Olesen and Henrik Keiding, Danida Forest Seed Centre, are thanked for their comments on the manuscript. Furthermore, Kirsten Tind and Anni Sloth are thanked for drawings and valuable technical assistance.

The study was financed by Danish Research Council for Developmental Studies (DANIDA), grants no. 104.Dan.8/469 and 104.Dan.8/402.

Author

Knud Tybirk. *Born 1960. Cand. scient. Aarhus University 1988. Since 1988 research associate and Ph.D. student financed by Danish Research Council for Development Studies (DANIDA) on a project entitled: Survival Strategies of Sahelian Woody Legumes in Relation to Management. Address: Botanical Institute, Aarhus University, 68 Nordlandsvej, DK-8240 Risskov, Denmark.*

1. INTRODUCTION

The last two decades have brought several dry periods in Sahel and the term desertification is found everywhere. However, desertification has not yet been proven in terms of detailed vegetation studies over large areas (Olsson 1985). The natural resilience of vegetation in areas with dynamic climatic fluctuations has to be kept in mind; the vegetation in such areas is well adapted to oscillating conditions, and if rainfall in Sahel during the coming decade is slightly above average levels, problems of regeneration of woody vegetation will depend mostly on management strategies in the area. Knowledge about legume trees is essential for making these management strategies.

The woody legumes (Leguminosae) in west African semi-arid savannas have great importance for the people and the ecosystem. Most dry zone legumes fix nitrogen, many serve as fodder for livestock, provide firewood, building material, resins, medicine or human food. The woody legumes are definitely the most important group of woody plants in these areas. However, little is known about their regeneration under "natural conditions", *i.e.* without being planted. Scattered information on seed dispersal, predation, germination (often only in nurseries), vegetative propagation, growth of young plants, *etc.* can be found, but very few attempts have been made to correlate these results with the main "natural" constraints for regeneration of the savanna legumes: fire, drought, browsing and attack of seeds by insect predators.

When focusing on flowering, pollination, seed production and seed ripening even less information is available (Doran *et al.* 1983; Tybirk 1989). This book intends to give an overview of the strategies for regeneration under the given conditions (fires, climatic variations, browsing by livestock, *etc.*) of the common woody legumes found in West Africa in the Sahelian and dry Soudanean vegetation zones as defined by White (1983). The species treated here are the legumes included in "Trees and Shrubs of the Sahel" by Maydell (1986) with main focus on indigenous species, but some important introduced species which are more or less naturalized are also included. The survey includes 16 genera with 38 species; 7 species of the family Caesalpiniaceae, 5 species of Fabaceae and 26 species from Mimosaceae (17 of which are acacias). Some of the species have a wide range of distribution, penetrating into Sahara or into the Soudanean zone, but they are all found on favorable sites in the Sahelian zone and are important in the Sahelian ecosystem.

Botanical taxonomy mostly follows Hutchinson and Dalziel (1954–58): Flora of Tropical West Africa, but for African acacias Ross (1979) is followed. Authors of scientific plant names can be found in chapter 7. Zoological taxonomy is rather diffuse as no general revisions of insect groups are available; when known, author names to scientific insect names are also included in chapter 7.

Regeneration will be viewed as much as possible as a natural process, without the involvement of humans. However, only few areas in west African savannas are not managed in some way by humans, but this does not prevent one from imagining how regeneration would be if the vegetation was left alone or managed in another way by man. Chapter 7 gives a short description of how to propagate the different tree species.

2. SEED DISPERSAL

After the development and maturation of the fruit and seeds, the next — and often crucial — step in the life-cycle of the plant is dispersal of its offspring. The strategy must be to disperse as many seeds with sufficient reserves (and capable of resisting environmental stress) in as many and/or as favorable conditions for germination and establishment as possible. This problem has been solved in many different ways depending on the species and their environment. Dispersal mechanisms must be related to the ancestry of the plant, but the legume family alone has evolved hundreds of different adaptations. Fruits of a legume species seem more readily modified evolutionarily than floral structures (Augspurger 1989).

As an introduction to the discussion of dispersal systems of Sahelian woody legumes, it is beneficial to discuss some selective advantages of seed dispersal in general.

Dispersal can be defined as transport of the diaspore away from the parent plant in time and space. The diaspore is the unit of the plant that is actually being dispersed (a single seed, seed and aril, a fruit, a group of fruits, *etc.*) and can often be determined on the basis of morphology of the fruit/seed. The term dispersal syndrome has been defined by O'Dowd and Gill (1986) as 'the non-random occurrence of combinations of diaspore traits related to the nature of the most probable dispersal agent.'

Most seeds are not dispersed over long distances but only a few meters to tens of meters (Venable and Brown 1988). Still, dispersal is a strong selective force, but also protection of seeds against unfavorable environment, germination and seedling establishment acts on the evolution of dispersal systems. Desert plants often have evolved characteristics that severely restrict dispersal in space, but instead evolve seed dormancy to disperse in time (Ellner and Shmida 1981, Renner 1987). Several hypotheses have been proposed to describe the selective advantages of dispersal of seeds (Howe and Smallwood 1982):

1. Escape hypothesis — Density of propagules is typically highest near the parent plant and conspecific competition and density dependent predation will therefore be highest in the vicinity of the parent. Thus, dispersal of seeds away from vicinity of the parent plant will be more successful.

2. Colonization hypothesis — In a changing habitat dispersal in time and space as far away as possible enables a parent to take advantage of incompetitive environments as they open. Short lived plants with quick reproduction, seed dormancy, self-fertilization, wind or animal dispersal will have the advantage.

3. Directed dispersal hypothesis — If rare micro habitats, well suited for establishment (*e.g.* specific soil conditions, termite mounds, *etc.*), are crucial for the seedling, directed non-random dispersal of seeds is essential.

In a given case of seed dispersal all three hypotheses may explain part of its selective advantage and it is their relative importance which determines the dispersal syndrome.

Different syndromes have different geographical and seasonal distribution. In tropical moist forests at least 50% and often 75% of the tree species produce fleshy fruits which are adapted for consumption by birds or mammals, while wind dispersed species are relatively more common in drier, more open, habitats (Howe and Smallwood 1982, Augspurger 1989). Wind dispersed fruits also have a tendency to be produced during the dry season, whereas animal dispersed species tend to be fruiting during the wettest months. The phenology is also correlated to conditions for germination and establishment (Garwood cited in Howe and Smallwood 1982).

Dispersal syndromes in Sahelian woody legumes

Legumes here considered are trees and shrubs adapted to semi-arid environment, a feature which should favor wind dispersal, which is found in about 50% of the species, the rest being mainly dispersed by the large mammals (wild and/or domestic) found in the Sahel. The large natural ungulate fauna of African savannas, some of which is extinct or severely reduced in West Africa today, has indeed favored adaptations for ungulate dispersal. A few species may also be adapted to water dispersal and bird dispersal. The following attempt to determine the natural dispersal syndrome of the woody legumes in Sahel is based on morphology of the diaspores, personal observations, and literature. Confirmed observations of dispersal mechanisms have been reported only in a few *Acacia* species and only the most probable or important will be described.

Wind dispersal. — Adaptations to wind dispersal are found in all three subfamilies of the Leguminosae and these diaspores are mostly indehiscent and single seeded (Augspurger 1989). However, of the listed species (Table 1) only a few have single-seeded indehiscent diaspores. This feature may be due to a secondary benefit from animal dispersal in Sahel.

In general, adaptation to wind dispersal is manifested as an increase in the surface/weight ratio (Schmidt 1988) and efficiency of wind dispersal depends on height of release, velocity of winds, number, size and location of seeds, segmentation, shape and size, of flattened structures such as wings, *etc.* (Schmidt 1988, Augspurger 1989). Wind dispersed diaspores are found more or less evenly distributed in the direction of the prevailing winds and not far from the parent plant. However, during three months of the rainy season of 1974 in west Africa there were 176 storms travelling distances of 580–800 kms with mean speeds of 50–61 km/hour (Hayward and Oguntoyinbo 1987). Such frequent and strong storms may carry diaspores for quite long distances and may be a strong selective force for the evolution of dispersal mechanisms.

In the case of Sahelian woody legumes the most common wind dispersed diaspores are 1) hemi-legumes which are dehisced papery pods separated in two halfs each with few–many flattened seeds attached, or 2) samaras which are whole or segmented flat papery fruits containing 1– few flat seeds (Figure 1A).

Actual observations of wind dispersal in the woody legumes of Sahel are rare. To my knowledge *Acacia mellifera* is the only species in which wind dispersal has been documented (Schmidt 1988, Tolsma 1989), but many authors (*e.g.* Buchwald 1895, Coe and Coe 1987, Augspurger 1989) agree that wind must be the most likely mode of dispersal of some of the species in Table 1. The rest are placed in this group because of similar pod morphologies.

The pods of these species (except *Mimosa pigra*) are probably also consumed by large herbivorous animals either from the tree or from the ground. However, survival rate after chewing and ruminating by large herbivores was much lower for flat seeds of six *Acacia* species with dehiscent pods than for rounded seeds of species with indehiscent pods (Coe and Coe 1987). Schmidt (1988) showed that 98.4% of the seeds of *A. mellifera* were digested during passage through a goat. Thus, animal dispersal is possible but it is not very likely a successful dispersal agent for the species mentioned in Table 1.

Acacia ehrenbergiana and *A. seyal* have been grouped here although their pods are not as flat and papery as those of the rest of the acacias all of which have hooked thorns; these two species with straight thorns will be mentioned again in the section on animal dispersal which

---

Table 1. List of wind dispersed woody legumes in Sahel.

Dehiscent hemi-legumes

> *Acacia ataxacantha, A. dudgeoni, A. ehrenbergiana, A. gourmaensis, A. laeta, A. macrostachya, A. macrothyrsa, A. mellifera, A. pennata, A. polyacantha, A. senegal, A. seyal, Albizia chevalieri, A. lebbeck, Cassia siamea, Leucaena leucocephala*

Entire samaras

> *Dalbergia melanoxylon, Pterocarpus erinaceous, P. lucens*

Segmented samaras

> *Entada africana, Mimosa pigra*

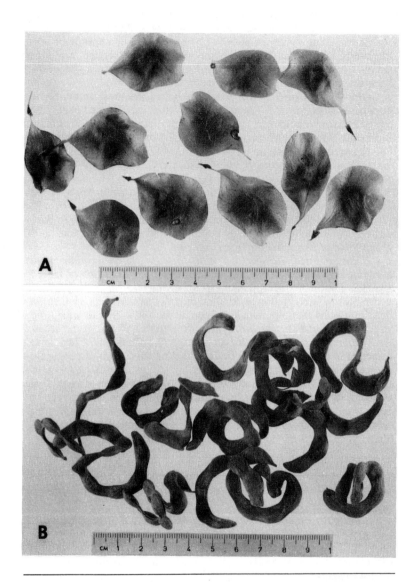

Figure 1A. The samara of *Pterocarpus lucens* is typically wind dispersed. **1B.** The pods of *Acacia tortilis* are eagerly sought by wild and domesticated ungulates and the seeds are dispersed by ungulates.

may be their main mode of dispersal. Buchwald (1895) also mentioned *Dicrostachys nutans* (= *D. cinerea*) and *Acacia spirocarpa* (= *A. tortilis* ssp. *spirocarpa*) as tumbleweeds dispersed in strong winds, either as a whole infructescence in *D. cinerea* or as a whole pod in *A. tortilis*. This possibility exists, but both species are more likely animal dispersed. The same author also mentioned that *A. arabica* (= *A. nilotica* probably ssp. *tomentosa*) is spread by wind after segmentation into "kreisscheiben-förmige Hülsenglieder" (= samaras). Truly, the pods of this species often separate into single seeded segments which may be dispersed by strong winds, but it is more likely that they are dispersed by animals or water.

Animal dispersal. — Endozoochory, dispersal through the digestive tract of animals, will give a more patchy distribution of seeds than wind dispersal and often over longer distances (Harper 1977, Fenner 1985). The behavior of animals, such as territoriality or seasonal wandering, fenced or free domesticated animals, and the time used for digestion determines the distribution of seeds. Endozoochory has been documented in a number of Sahelian legumes or related species through the finding of viable seeds in feces (*e.g.* Burtt 1929, Leistner 1961, Lamprey 1967, Wickens 1969, Gwynne 1969, Cheema and Quadir 1973, Halevy 1974, Lamprey *et al.* 1974, Jarman 1976, Mooney *et al.* 1977, Harvey 1981, Hopkins 1983, Sabiiti and Wein 1987, Coe and Coe 1987, Schmidt 1988, Hauser unpublished report, Tolsma 1989). Documented cases of endozoochory exist for *Acacia albida, A. mellifera, A. nilotica, A. sieberiana, A. senegal, A. seyal, A. tortilis, Cassia sieberiana*, and *Tamarindus indica*, but many other species with similar pod and seed morphology are eaten by animals. Although viable seeds have not been documented in the feces of the assumed dispersal agents, they may tentavely be considered endozoochorous plant species (Table 2). Most endozoochorous species are dispersed by large ungulates, either wild or domesticated, but a few species are dispersed or eaten by birds, monkeys, and humans. The fruits of endozoochorous species should be rewarding for the consumer, for instance have fleshy mesocarp, strong smell, and/or be available in large amounts in periods of food scarcity (Figure 1B; Gwynne 1969, Pijl 1982, Coe and Coe 1987). Furthermore, the seeds must be able to withstand chewing and passage through the digestive tract. A reward for the consumer may be that some of the seeds are digested at the same time as some harder seeds are dispersed as in the case of bird dispersal of *Acacia seyal* (Schmidt 1988). Ahmed El Houri (1986) observed tooth marks in many *A. tortilis* seeds in feces, suggesting that the softest seeds had been consumed.

Defecation by ungulates will often happen in disturbed favorable sites, *e.g.* near watering points, along trails, or in the shade of a tree (Leistner 1961). Goats and sheep are superior to cattle and camels as

---

Table 2. List of animal dispersed woody legumes in Sahel.

Typically ungulate dispersed
 Acacia albida, A. nilotica, A. sieberiana, A. tortilis, Bauhinia rufescens, Cassia sieberiana, Dichrostachys cinerea, Parkia biglobosa, Piliostigma reticulatum, P. thonningii, Prosopis africana, P. juliflora, Tamarindus indica.

Possibly ungulate dispersed
 Acacia ataxacantha, A. dudgeoni, A. ehrenbergiana, A. gourmaensis, A. laeta, A. macrostachya, A. macrothyrsa, A. mellifera, A. pennata, A. polyacantha, A. senegal, A. seyal;., Albizia chevalieri, A. lebbeck, Cassia siamea, Dalbergia melanoxylon, Leucaena leucocephala, Mimosa pigra, Parkinsonia aculeata, Pterocarpus erinaceous, P. lucens.

Bird dispersed
 Acacia ehrenbergiana, A. seyal, Erythrina senegalensis Parkia biglobosa, Tamarindus indica.

Primate dispersed
 Acacia tortilis, Dichrostachys cinerea, Parkia biglobosa, Parkinsonia aculeata, Prosopis juliflora, Tamarindus indica.

---

dispersal agents of *A. tortilis* because their small round fecal pellets are dispersed by wind and/or rain to favorable sites (Ahmed El Houri 1986). Secondary dispersal by dung beetles may be of importance to reduce conspecific competition or predation of seeds in the feces (Coughenour and Detling 1986, Janzen 1986, Hauser unpublished report). Germination is favored by shade and high soil moisture and the feces affect the germination rate directly as well as water and nutrient balance of the establishing seedling (Preece 1971, Leistner 1961). However, several authors mention disadvantages of feces for germination, such as high ammonia concentration, competition (Janzen 1971, Pijl 1982, Wickens 1969).

Possible epizooic (the diaspore adhering to the animal) dispersal among woody legumes in Sahel includes only *Pterocarpus erinaceous* and *Mimosa pigra*. Tolsma (1989) suggests that *Dichrostachys cinerea* pods may be dispersed between the hooves of cattle. Buchwald (1895) does not consider it likely that straight bristles without hooks on the samara of *P. erinaceous* will attach to the fur of animals; he sees the bristles as a defense against endozoochory and belives the species is wind dispersed. The dense bristles must in some cases be able to become attached to animals capable of carrying the seeds for quite long distances, up to several kilometers. Buchwald (1895) mentioned the densely bristled segmenting pod of *Mimosa pigra* as a possible adaptation to epizooic dispersal. *Mimosa aperata* (= *M. pigra*) is water dispersed when growing along rivers in the Amazon basin, but animal dispersed in dryer habitats when growing along animal trails (Ridley 1930, Pijl 1982). Probably wind can also play a minor role, but endozooic dispersal seems likely although cattle avoid the species for browsing. This introduced species has spread over large areas of semi-arid Africa and it has an efficient dispersal systems adapted for several habitats (Maydell 1986).

Water dispersal. — Dispersal by water is a well known phenomenon along coasts, swamps and permanent rivers in humid areas (Ridley 1930, Pijl 1982), but not a very common feature in dry Sahelian climate where nearly all water courses are temporary. The water current in these may, however, be very strong during the rainy season and if dry pods are dropped at that time, long distance water dispersal may occur. One should keep in mind that water dispersal is restricted in direction and always down stream; therefore, other agents are always needed for upstream dispersal. Fruits must have a high surface/weight ratio, be water repellent, and the seeds must be able to withstand water-logging for several days without rotting or germinating. Seeds of water dispersed plants will often land on favorable sites for germination such as disturbed, humid, nutrient rich soil.

Dispersal by water is difficult to observe in nature, but indications of it can be found in several ways. Present distribution of *Acacia gerrardii* in Israel is a reflection of fluvial draining systems during the mid-Pleistocene (Halevy and Orshan 1972) and the dispersal of introduced *A. melanoxylon* in South Africa for 32 km along a stream during 13 years is a strong indication of water dispersal (Milton and Hall 1981). None of these species have specific adaptations for this form of dispersal, but adaptations for endozooic dispersal and soil requirements may also favor water dispersal. The distribution and pod morphology of some Sahelian species suggest water dispersal (Table 3).

Water dispersal in *Acacia albida* was not very likely because its ripe pods sink after three days in water (Wickens 1969), but pods may float after one week in water and seeds may survive water-logging for 16 days (Hauser, unpublished report) which indicates that long-distance water dispersal is not only possible, but very likely in populations along water courses. Also in the case of *A. nilotica*, which is frequently found along rivers or in water-logged depressions, the dry pods may float in water for some time and germinating seeds were found in massive concentrations of pods on the shore of a temporary lake in Senegal

---

Table 3. List of water dispersed woody legumes in Sahel.

Possibly water dispersed species
  Acacia albida, A. nilotica, A. sieberiana, Piliostigma reticulatum, P. thonningii

---

Figure 2A and B. Massive concentrations of pods of *Acacia nilotica* ssp. *tomentosa* on the shores of a temporary lake in Senegal is an example of water dispersal.

(Figure 2). Water dispersal may also happen from time to time in many other species during heavy rains on soils where runoff is large.

Possibilities for dispersal
From the examples described above it is clear that Sahelian woody legumes have several different ways of dispersal depending on local conditions. They can be listed as primary and secondary ways of dispersal for the individual species (Table 4).

Table 4. Dispersal of Sahelian woody legumes.

Species	Wind		Animals			Water
	hemi-legume	samara	ungu-late	bird	pri-mate	
Acacia albida			p			s
Acacia ataxacantha	p		s			
Acacia dudgeoni	p		s			
Acacia ehrenbergiana	s		s	p		
Acacia gourmaensis	p		s			
Acacia laeta	p		s			
Acacia macrostachya	p		s			
Acacia macrothyrsa	p		s			
Acacia mellifera	p		s			
Acacia nilotica			p			s
Acacia pennata	p		s			
Acacia polyacantha	p		s			
Acacia senegal	p		s			
Acacia seyal	s		s	p		
Acacia sieberiana			p			s
Acacia tortilis			p		s	
Albizia chevalieri	p		s			
Albizia lebbeck	p		s			
Bauhinia rufescens			p			
Cassia siamea	p		s			s
Cassia sieberiana			p			
Dalbergia melanoxylon		p	s			
Dicrostachys cinerea			p		s	
Entada africana		p				
Erythrina senegalensis				p		
Leucaena leucocephala	p		s			
Mimosa pigra		s	p			s
Parkia biglobosa			p	p	p	
Parkinsonia aculeata			s		s	
Piliostigma reticulatum			p			s
Piliostigma thonningii			p			s
Prosopis africana			p			
Prosopis juliflora			p		s	
Pterocarpus erinaceous		p	s			
Pterocarpus lucens		p	s			
Tamarindus indica			p	s	s	

p = primary dispersal agent
s = secondary dispersal agent

3. SEED PREDATION

Probably everyone working with woody legumes in the tropics has found seeds with clear signs of insect attack: *e.g.* rounded exit holes in an empty seed shell. This chapter will summarize knowledge about these seedborers and their host plants in Sahel, infestation rates, relations to fire, dispersal, and germination.

Seed predation is here defined as: consumption of seeds by animals killing or severely reducing viability of the seed. This is a clearcut definition in most cases, but it may be difficult to separate predation from dispersal in some cases because some dispersal agents may kill or reduce viability of the seeds. Seed predation is often thought of as insects eating mature seeds, but in the case of Sahelian woody legumes, ungulates acting as dispersal agents of mature seeds may at the same time predate on developing seeds by eating green pods on the trees.

Coe and Coe (1987) mentioned that seeds of acacias only mature out of reach of ungulates. They assumed that ungulates eat most of the green pods within reach during development. To reduce the costs of this predation pods of acacias reach their full length before the seeds swell and mature. However, Vervet Monkeys have been observed feeding on swollen green seeds and pods of *A. tortilis* (Wrangham and Waterman 1981). Consumption of developing seeds reduces the potential seed crop of certain species of acacias in areas with large ungulate populations (Pellew and Southgate 1984).

It may also be difficult to separate predation from dispersal when seeds in the feces of an ungulate germinate, but die within few days due to high ammonia concentration. Also the consumption of dehiscent pods by ungulates will in most cases be predation even if the seeds are mature (see page 7).

Insects on legume seeds

Legume seeds are attacked by a variety of insects in the orders Coleoptera (beetles), Hemiptera (bugs), Lepidoptera (butterflies and moths), and Hymenoptera (wasps and ants). The beetle family Bruchidae has by far the largest ecological and economical importance and this section will focus on interactions between bruchids and woody legumes in Sahel. The impact of bruchid predation on seeds has been described as the primary factor regulating *Acacia* populations in Africa (Lamprey *et al.* 1974, Karschon 1975).

Bruchidae is a beetle family adapted to feed on legume seeds and it is closely related to leaf beetles (Chrysomelidae) and long-horned wood borers (Cerambycidae). The superficial resemblance to snout beetles (Curculionidae) has led to the misleading common name "seed weevils"; instead the common name "bean beetle" should be used. The cosmopolitan family Bruchidae includes about 1300 species in six subfamilies (80% in subfamily Bruchinae) and 56 genera of which 10 are found in the Old World (Johnson 1981). The African species have not been revised taxonomically recently and some confusion exists regarding the identity of species. Knowledge about distribution of species, their relations to host plants, and general ecology is sparse (Southgate 1978, 1979, Tolsma 1989).

Ecology and biology of bruchids. — Bruchids are not exclusively linked to legumes, but 84% of the species attack legume seeds. Their larvae are known to live only in seeds while the adult beetles feed on pollen and nectar. Based on their biology the bruchids can be separated into two groups: 1) bruchids attacking developing pods (mostly of Caesalpinaceae and Mimosaceae) and 2) bruchids attacking ripe pods or seeds (mostly of Fabaceae). Based on their life cycle they can be separated into bruchids being 1) univoltine with one generation per year or 2) multivoltine with more generations per year, the latter being the most common (Southgate 1981, 1983).

Lifecycle of bruchids. — Adult female bruchids seek a developing pod of a woody legume after mating and lay their eggs on the pod surface within 30–100 days (30–60 days in *Acacia tortilis* according to Pellew and Southgate 1984) after pollination of the tree. After 5–10 days the larvae penetrate the pod and enter the seeds by rasping away material with their mouthparts (Southgate 1983).

Only one larva per seed is present if the seeds are small, whereas more larvae per seed may be found in species with large seeds (indehiscent species). In the seed the larva moults four times and pupates; after 3–12 weeks, depending on species and the weather conditions, the adult beetle emerges leaving the characteristic bruchid hole in the seed (Figure 3). The generation time in *Bruchus uberatus* (= *B. baudoni*) has been shown to be 39 days in humid experimental conditions (Peake 1952), but 200–250 days in Botswana where hatching of the beetle coincides with the occurrence of young pods (Tolsma 1989). Usually, the seed cotyledons and the embryo have been eaten or damaged leaving the seed non viable (Johnson 1981, Southgate 1983, Coe and Coe 1987).

The adult beetle is able to diapause when fresh pods or seeds are not available for oviposition. Fertility and life time of the adult beetle can be extended if it feeds on pollen and nectar (Johnson 1981). However, survival of adult beetles is independent of supply of water and/or sugar, fresh flowers, *etc.* (Tolsma 1989) Several beetles of *Bruchidius uberatus* survived 20–30 days in the laboratory. Some species spend the dry season in pupal cocoons in the soil (Donahaye *et al.* 1966). In West Africa the emergence of the adult bruchid, depending on the species, takes place from October to July after ripening of seeds of the hosts (Nongonierma 1978).

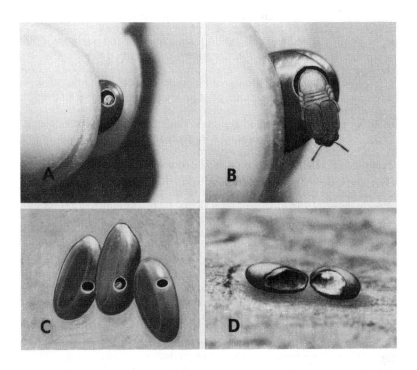

Figure 3. The adult bruchid beetle emerges from the seed of *Acacia albida* (**A, B**) and the empty seed shell is left with the characteristic exit hole. (**C, D**). Photo: Thure Hauser.

Many species prefer developing pods for oviposition, but where seeds are stored, most bruchids are capable of producing several generations which may destroy all the seeds in the store. Some species have been shown to emerge from seeds of *A. nilotica* after several years of storage, but most emerge after 2–3 months after seed collection (Nongonierma 1978, Tolsma 1989).

Host specificity. — Legumes have evolved several defence mechanisms against seed predation. Hard seeded legumes have two strategies: 1) develop many small seeds with reduced chemical defense (predator

satiation strategy) or 2) develop few larger seeds maintaining the
chemical defence (Janzen 1969). Janzen listed 31 traits in Leguminosae
which reduce seed infestation, but these defensive traits seem to be
effective against one or few species of bruchids, rarely all.

Fruit morphology, size, and hardness of the seed coat are
important features for the plant in selection against certain beetles, but
the chemistry of seeds is more important (Janzen 1969, 1971). Low
molecular weight nitrogenous compounds, which are toxic to most
animals, are found in the seeds of legumes. Bruchids have adapted to
feed on these. Perhaps bruchids even shifted onto legume seeds because
of these compounds, and at the same time the bruchids have lost the
ability to break down proteins. These traits have led to a certain degree
of host specificity, the legume evolving toxic compounds is countered by
evolution of bruchids feeding on these compounds.

The bruchids are mostly genus specific and have a narrow host
preference on closely related species (Johnson 1981). Of 61 species of
Acacia described in Flora of Tropical East Africa (Brenan 1959) only 14
are known to host more than one species of bruchids (Southgate 1978).
Common legume species with wide distributions seem to have more seed
predators (Ross 1979, Tonder cited in Coe and Coe 1987), but host
specificity was very variable in a study by Tolsma (1989). A single seed
may contain one or several bruchid species. Species with large seeds
have greater probability of hosting more species or individuals and some
bruchid species are known to feed on more than a single (small) seed to
complete life-cycle (Coe and Coe 1987, Skaife 1926).

Predators of Sahelian woody legumes

Bruchid seed predators. — The list of predators in Table 5 (page 18) has
been compiled from literature and is not a complete list, but the best
available at the moment (see also Tolsma 1989). The taxonomy of many
of the species is unclear and some of the names used below may be
synonyms. A revision of African bruchids is badly needed. Table 5 is
based on research in all parts of Africa and Israel in more or less similar
ecological conditions and shows Sahelian legume species hosting a
number of species and genera of (mostly) bruchids.

Note that some confusion exists between the genera *Bruchus* and
Bruchidius; the latter was described early this century (Southgate 1983).
It should also be noted that the genera *Acanthoscelides, Amblycerus,
Algarobius, Caryedes, Mimosestes, Rhipobruchus,* and *Scutobruchus* are
New World genera known to infest their hosts in their native range. No
information exists on introduction of these species with their host plant to
Sahel.

Non-bruchid seed predators. — Hemiptera (bugs), Lepidoptera
(butterflies and moths), Hymenoptera (wasps and ants) and other
Coleoptera (beetles) than bruchids may also attack seeds of woody

legumes but very few records exist of identified host-predator relationships and this field seems even more unexplored than bruchid infestation.

The study by Ernst *et al.* (1989) mentions that the phytophagous wasp *Oedante* sp. predates on seeds of *Acacia tortilis* and Southgate (1983) mentions that the nymphs of the Hemipteran genus *Nemasus* attack *A. tortilis* ssp. *raddiana* seeds and that the phytophagous Hymenopteran genus *Bruchophagus* feeds on seeds of *Acacia* spp. Also the beetle families Anobiidae and Curculionidae have been mentioned as granivores of *Acacia albida* and *A. macrostachya*. Curculionidae has been described as the "ecological homologue" to bruchids on Australian acacias (New 1983). The Buffalo Tree Hopper infests 17–83% of seeds of *A. senegal* in Pakistan (Cheema and Quadir 1973) and also rodents, millipedes, and microorganisms predate on seeds of *A. senegal* in Sudan (Seif el Din and Obeid 1971a). Ants may also attack hard legume seeds (Gissel, pers. comm.).

Bruchid parasites or predators. — Bruchids have natural enemies, but only little is known about their biology and impact on bruchid populations. Most parasites belong to Hymenoptera and attack eggs, larval and pupal stages, but also mites (Acarina) and flies (Diptera) attack bruchids (Southgate 1979).

Several species of the hymenopteran genus *Uscana* are known to attack up to 25% of the eggs in the field of *Bruchus, Bruchidius, Callosobruchus, Acanthoscelides* (Skaife 1926), and *Caryedon* and they seem to be host specific. The parasitic genus *Bruchobius* is known to feed on larvae and pupae of *Bruchidius* species associated with acacias (Southgate 1983). The mite *Pynotes boylei* is known to predate on adult bruchids (Southgate 1983). Johnson (1983) mentions several parasites of bruchids associated with American *Prosopis* spp. attacking 0–25% of bruchids.

The following list (Table 6, page 20) gives an impression of the sparse information from literature on host-predator-parasite relationships on legumes found in Sahel.

Much more research is needed to clarify the taxonomy and ecology of these pests, before clarifying the extent and potential of such natural limitators of bruchid populations. Already Skaife (1926) mentioned that some wasps are easily reared in the laboratory and might be a potential way of biological control of natural bruchid populations.

Extent of seed predation
The infestation ratio of seeds is variable among the different genera, species and individuals within and between different seasons. Written evidences of infestation vary from 0-100%, but these numbers do not only reflect biological and climatic differences, but also the different methods of testing seeds (Table 7, page 21) and the time of collection.

Table 5: List of host and bruchid seed predators.

Tree species	Bruchid species
Acacia albida	*Bruchus natalensis, Bruchidius auratopubens, B. aurivillii, B. cadei, B. platypennis, B. pygidiopictus, B. silaceous, B. sp. near rufulus, Bruchidius sp., Caryedon excavatus, Pachymerous longus, P. pallidus, 2 Tuberculobruchus spp.*
Acacia ataxacantha	*Bruchidius cadenati, B. silaceus, B. submaculatus, Bruchidius sp., Caryedon mauritanicus, C. pallidus, Tuberculobruchus sp.*
Acacia dudgeoni	*Bruchidius cadenati, B. silaceus, B. tougourensis, Caryedon mauritanicus,*
Acacia ehrenbergiana	*Bruchidius sahelicus, B. uberatus, Caryedon sahelicus,*
Acacia gourmaensis	*Bruchidius cadenati, B. submaculatus, Caryedon mauritanicus*
Acacia laeta	*Bruchidius cadenati*
Acacia macrostachya	*Bruchidius auratopubens, B. aurivillii, Bruchidius cadenati, B. pennatae, B. silaceus, B. submaculatus, B. tougourensis, Caryedon mauritanicus, Tuberculobruchus natalensis*
Acacia macrothyrsa	*Bruchidius senegalensis*
Acacia mellifera	*Bruchidius sp., Caryedon pallidus*
Acacia nilotica	*Bruchus maculatus, B. silaceus, Bruchidius analis, B. albonotatus, B. albosparsus, B. bedfordi, B. baudoni, B. centromaculatus, B. luteopygus, B. quadrimaculatus, B. senegalensis, B. sieberianae, B. submaculatus, B. theobromae, B. uberatus, Bruchidius sp., Caryedon capicola, C. albonotatum, C. interstinctus, Pachymerus cassiae, P. longus, P. pallidus, Tuberculubruchus natalensis*
Acacia pennata	*Bruchidius cadenati, B. pennatae, Caryedon mauritanicus*
Acacia polyacantha	*Bruchidius cadenati, B. campylacanthae, B. senegalensis, B. silaceus, Caryedon mauritanicus*
Acacia senegal	*Bruchus centromaculatus, B. subuniformis, B. silaceus, Bruchidius albosparsus, B. cadenati, B. pennatae, B. petechialis, B. sahelicus, B. senegalensis , B. submaculatus, B. uberatus, Bruchidius sp., Caryedon mauritanicus, C. longispinosa, C. pallidus, C. sahelicus, Pachymerus pallidus, Tuberculubruchus natalensis*

(continued...)

Table 5: List of host and bruchid seed predators (continued)

Tree species	Bruchid species
Acacia seyal	*Bruchus elnairensis, Bruchidius aurivillii, B. sahelicus, B. sieberianae, B. voltaicus, Caryedon capicola, C. excavatus, C. mauritanicus , C. sahelicus*
Acacia sieberiana	*Bruchidius baudoni, B. natalensis, B. centromaculatus, B. luteopygus, B. platypennis, B. sahelicus, B. sieberianae, B. uberatus, Bruchidius sp., Caryedon albonotatum, C. interstinctus, Tuberculubruchus natalensis*
Acacia tortilis	*Bruchus elnairensis, B. albonotatus, Bruchidius albonotatus, B. albosparsus, B. aurivillii, B. cadei, B. cadenati, B. centromaculatus, B. latevalvus, B. petechialis, B. rubicundus, B. sahelicus, B. silaceus, B. sinaitus, B. spadicus, B. uberatus, Caryedon capicola, Caryedon gonagra, C. longispinosa, C. sahelicus, Caryedon serratus, Caryedon sp. Spermophagus rufonotatus*
Albizia lebbeck	*Bruchus submaculatus, B. tougourensis*
Bauhinia rufescens	*Caryedon gonagra, C. cassieae, C. serratus*
Cassia sieberiana	*Caryedon gonagra, C. serratus, C. cassieae*
Dicrostachys cinerea	*Bruchidius petechialis, B. centromaculatus, B. dichrostachydis, B. petechialis, B. securiger, Bruchidius spp., Spermophagus densepubens, S. rufonotatus*
Parkinsonia aculeata	*Caryedes germanii, Mimosestes amicus*
Piliostigma reticulatum	*Caryedon serratus*
Piliostigma thonningii	*Caryedon serratus*
Prosopis africana	*Caryedon cassieae*
Prosopis juliflora	*Acanthoscelides* sp., *Amblycerus epsilon, A. prosopis, Algarobius bottimori, Caryedon* sp., *Mimosestes amicus, M. insularis, M. protractus, Rhipobruchus prosopis, R. psephenopygus, Scutobruchus ceratioborus*
Tamarindus indica	*Caryedon serratus*

References: Decelle 1951, Peake 1952, Donahaye *et al.* 1966, Prevett 1965, 1966, 1967, Wickens 1969, Howe 1972, Center and Johnson 1974, Halevy 1974, Lamprey *et al.* 1974, Karschon 1975, Southgate 1975, Kingsolver *et al.* 1977, Belinski and Kugler 1978, Nongonierma 1978, Varaigne-Labeyrie and Labeyrie 1981, Johnson 1983, Pellew and Southgate 1984, Robert, 1985, Tonder 1985, Ernst *et al.* 1989, Tolsma 1989, Tybirk unpublished.

Table 6. Host -predator-parasite relationships on Sahelian woody legumes.

Host	Predator	Parasite	Reference
Acacia albida	*Bruchidae spp.*	*Eupelidae spp.*	1
Acacia ataxacantha	*Bruchidae spp.*	*Eupelidae spp.*	1
Acacia macrostachya	*Bruchidae spp.*	*Eupelidae spp.*	1
Acacia nilotica	*Bruchidius albosparsus*	*Entedon apionidis*	2
Acacia nilotica	*Bruchidius albosparsus*	*Senegalella leguminosae*	2
Acacia nilotica	*Bruchidae spp.*	*Eupelidae spp.*	1
Acacia senegal	*Bruchus baudoni*	*Erytoma bruchocida*	2
Acacia seyal	*Bruchidius albosparsus*	*Erytoma sp.*	2
Acacia tortilis	*Bruchidae spp.*	*Eupelidae spp.*	1
A.cacia tortilis	*Caryedon gonagra , Bruchus albosparsus*	*Oedaule stringifrons*	3
Acacia tortilis	*Caryedon gonagra, Bruchus albosparsus*	*Anisopteromalus calandrea*	3
Acacia tortilis	*Caryedon gonagra, Bruchus albosparsus*	*Bruchocida vuilleti*	3
Acacia tortilis	*Caryedon gonagra, Bruchus albosparsus*	*Habrobracon brevicornis*	3
Acacia tortilis	*Caryedon gonagra*	*Exoprosopa minops*	3
Acacia tortilis ssp. raddiana	*Bruchus albosparsus*	*Oedaule italica*	2
Acacia tortilis ssp. raddiana	*Bruchidae spp.*	*Metriocharis silvestris*	4
Acacia tortilis ssp. raddiana	*Bruchidae spp.*	*Eupelmus sp.*	4
Piliostigma reticulatum	*Caryedon gonagra*	*Bracon kirkpatricki*	5
Piliostigma thonningii	*Caryedon gonagra*	*Bracon kirkpatricki*	5
Tamarindus indica	*Caryedon gonagra*	*Bracon kirkpatricki*	5

References: 1. Nongonierma 1978; 2. Luca 1965; 3. Donahaye *et al.* 1966; 4. Ernst *et al.* 1989; 5. Prevett 1966.

Table 8 (page 22) gives percentages of seed predation found in literature. The methods used for determination of infestation may be any of the above mentioned (Table 7), but mostly it has been scored by presence of bruchid exit holes (method 1 and 2).

Except for *Dichrostachys cinerea*, only African acacias have been investigated and mostly the older pods or stored seeds have higher rates of infestation, but the variation is large. This is indeed confirmed

Table 7. Methods of estimating seed predation

1. Seeds collected from fresh pods and examined externally and scored as infested by presence of exit holes. The most widespread and simple method.

2. In stored seeds the beetles have had the opportunity of completing the life cycle of beetles not seen in 1). Scored as infested by presence of exit holes (Lamprey *et al.* 1974, Halevy 1974, Sabiiti and Wein 1987).

3. X-raying seeds has often revealed higher percentages, as also developing larvae and dead pupae can be seen by this technique (Pellew and Southgate 1984, Tybirk pers. obs.).

4. The seeds may be dissected and the ratio of viable embryos can be determined. Non-viable embryos may be taken as a measure of infestation.

5. Infestation has also been tested by floating of seeds. Infested seeds float (Peake 1952).

by the data presented by Ernst *et al.* (1989) who found bruchid predation on *A. tortilis* ssp. *heteracantha* from six trees varying from 37.0–81.8% in 1985, 9.5–23.7% in 1986 and 17.5–57.7% in 1987. One tree was found to vary from 10.5–69.7% infestation during these three years. In addition, chalcid wasps had attacked 0–2.2% of the seeds in 1985, 1.0–3.8% in 1986 and 9.6–12.0% in 1987. Also Pellew and Southgate (1984) and Lamprey *et al.* (1974) found large variation in predation percentages in the same stand of *A. tortilis* ssp. *spirocarpa* in Tanzania between different years. Also Tolsma (1989) found large variation in predation percentages on predation of *A. nilotica* seeds in Botswana.

Ecological impact of seed predation
The size of the viable crops of legume seeds in Sahel is severely reduced each year by seed predation. The impact of the seed predators on the crop is, however, controlled by a series of factors, especially seed dispersal, passage of seeds through ungulates, fire, and germination.

Effects of dispersal. — Dispersal of seeds in space is believed to be the most important strategy to reduce impact of bruchids (Janzen 1969, Coe and Coe 1987, Schmidt 1988, see also chapter 1). The density of seeds is higher under parent trees and the risk of a predator finding the crop will be higher there. Especially for trees in which predator species are capable of re-infesting dry seeds, it is crucial to disperse the seeds to avoid this. Low ungulate pressure on African acacias will result in higher bruchid predation rates (Halevy 1974, Coe and Coe 1987). Probably the early dispersal in space before beetle attack plays the largest

Table 8. Seed predation by insects on selected woody legumes in percentages of investigated seeds.

Species	Area	Infested seeds %			Reference
		fresh pods	old pods	stored seeds	
Acacia albida	Zimbabwe	4.1			1
	Zambia	13.0		12.4	2
	Sudan		51		3
	W. Afr.			38	4
Acacia ataxacantha	W. Afr.			20	4
Acacia caffra	S. Afr.	21.4			5
Acacia dudgeoni	W. Afr.			7	4
Acacia ehrenbergiana,	W. Afr.			97	4
Acacia elatior	Kenya	29.0		71.0	1
		21.6			
Acacia erioloba			28.7		1
Acacia giraffae	S. Afr.	33	>50		6
Acacia gourmaensis	W. Afr.			24	4
Acacia laeta	W. Afr.			32	4
Acacia macrostachya	W. Afr.			41	4
Acacia macrothyrsa	W. Afr.			6	4
Acacia mellifera	Kenya	25–35			7
Acacia mellifera	Botswana	0.8–26.2			8
Acacia nilotica	Sudan	8		61	9
	Kenya	0			7
	Kenya			1	10
	W. Afr.			36–49	4
	Botswana	0–61			8
Acacia pennata	W. Afr.			13	4
Acacia polyacantha	W. Afr.			13	4
Acacia senegal	W. Afr.			15–38	4
Acacia seyal	W. Afr.			12–16	4

continued

Table 8. (continued). Insect predation of selected woody legumes in percentages of investigated seeds.

Species	Country	Infested seeds %			Reference
		fresh pods	old pods	stored seeds	
Acacia sieberiana			35,0		1
	Uganda		96		11
	W. Afr.			34	4
Acacia tortilis	Kenya	28			12
	Tanzania	13			13
	Kenya	5–10			7
	Israel	11		2	14
	Botswana	11.9–78.8			8
Acacia tortilis ssp.					
- *heteracantha,*	S. afr.	7.6			1
- *heteracantha,*	Botswana	18.5–54.5			15
- *spirocarpa,*	Kenya	6.6			1
- *spirocarpa,*	Tanzania			95.6–99.6	16
Acacia tortilis ssp.					
- *spirocarpa*	Tanzania	35.6–84.4			17
- *spirocarpa*	Tanzanie	5.1–8.4			17
- *raddiana*	Israel		70–98		18
- *raddiana*	Israel	4; 26; 68		47.3	14
- *raddiana*	W. Afr.			70	4
Dichrostachys cinerea	Botswana	17–93.3			8

References: 1. Coe and Coe 1987; 2. Hauser unpublished report; 3. Wickens 1969; 4. Nongonierma 1978; 5. Ross 1965; 6. Leistner 1961; 7. Schmidt 1988; 8. Tolsma 1989; 9. Peake 1952; 10. Tybirk 1988; 11. Sabiiti and Wein 1987; 12. Coughenour and Detling 1986; 13. Jarman 1976; 14. Karschon 1975; 15. Ernst *et al.* 1989; 16. Lamprey *et al.* 1974; 17. Pellew and Southgate 1984; 18. Halevy 1974.

Fresh pods are collected on the tree or newly fallen pods. Old pods are collected from the ground, few weeks to several months old. Stored seeds may be from few months to several years of age.

role in nature. However, seed predation by bruchids on *Cassia biflora* in Costa Rica has been shown by Silander (1978) to be density independent as bruchids attack more pods but fewer seeds per pod at higher densities.

The passage of seeds of woody legumes through ungulates has probably some importance in avoiding predation. However, larger seeds of indehiscent species have space for more bruchid species than smaller wind dispersed seeds of dehiscent pods. Therefore, evolution of

Figure 4. The domestic ungulate populations in Sahel is of great importance for seed dispersal and seed predation.

characteristics adapted for mammal dispersal has also increased the susceptibility to seed beetle attack (Coe and Coe 1987), although Varaigne-Labeyrie and Labeyrie (1981) found no correlation between seed size and rate of bruchid infestation. Passage through the ungulate digestive system has been postulated to reduce the percentage of bruchid damage on seeds of several *Acacia* species by 1) killing eggs and larvae inside seeds, 2) enhancing germination and thereby reducing the damage by beetles, 3) dispersal in space, or 4) removing smell of seeds attracting the bruchids (Janzen 1969, Lamprey *et al.* 1974, Coe and Coe 1987, Figure 4). Hoffmann *et al.* (1989) found lower bruchid predation of seeds still on the tree compared to seeds on the ground, probably due to different age of the investigated samples as infestation was scored when exit holes were found.

The combination of bruchid attack and ingestion by ungulates will also kill many seeds either by simple digestion of seeds with exit holes or the weakened, softened, seeds will start germinating in the digestive tract and then eventually be digested. Hauser (unpublished report) found many more or less decayed *Acacia albida* seeds, 35% of all seeds, in old cattle feces in Zambia so the softening of seed coat during passage may kill many seeds. Schmidt (1988) found no *Acacia* seeds with exit holes of bruchids from goat droppings indicating that all infested seeds had been digested (n=2000). Removal of smell of seeds may play a role in nature even though Johnson (1981) reported bruchids laying eggs in seeds of indehiscent *Acacia* spp. in animal feces.

Post dispersal seed predation (*e.g.* rodents feeding on seeds in ungulate feces) may be of importance in some areas (Janzen 1986, Tolsma 1989), depending on the disperser, the seed and the predator.

Effects of fire. — Only one study has been made concerning the role of fire for infestation of bruchids, but the study indicated that this aspect may be very important in large parts of Sahel. The study was carried out in Uganda on seeds of *Acacia sieberiana* and showed that low intensity fires can selectively kill developing bruchid beetles inside the seed without killing the embryo (Sabiiti and Wein 1987). When the fires are more intense, the seeds attacked by bruchids are more apt to be killed by the fire, perhaps indicating that the entrance hole of the bruchid makes the seed vulnerable to fire as well as increasing water permeability (Figure 5).

This has important implications for management of savanna areas: if the pressure from grazing herbivores is high, less grass will act as fuel for fires, decreasing the intensity of fires, and many bruchid larvae in the seeds will survive. On the other hand, if few grazing herbivores are present, more grass will give a more intense fire, selectively killing more bruchids in the seeds, and also stimulating germination from the seed pool in question. This means that few grazers will give more seedlings

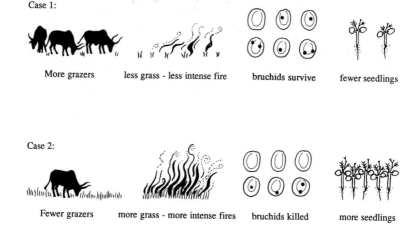

Case 1:

More grazers   less grass - less intense fire   bruchids survive   fewer seedlings

Case 2:

Fewer grazers   more grass - more intense fires   bruchids killed   more seedlings

Figure 5. The complex interactions between grazers, fuel (grass), bruchid development, germination and seedling establishment as it may work in a savanna in Uganda. See text.

of *A. sieberiana* with the observed fire intensities (Sabiiti and Wein 1987). The effects will be greatest on multivoltine bruchid species that are able to re-infest seeds in the soil (Tolsma 1989) and the timing of fires is also important.

Effects of germination. — The impact of bruchid infestation on germination has been discussed by Lamprey *et al.* (1974), Coe and Coe (1987), Ernst *et al.* (1989) and Hauser (unpublished report). If a bruchid consumes the embryo of the seed, no germination will occur. Skaife (1926) observed that 40% of the embryos of bruchid infested seeds in South Africa had been destroyed or damaged. If the bruchid larvae eat parts of the cotyledons, the seeds remain potentially viable and seeds with bruchid exit holes have been reported germinating (less than 10%). Southgate (1978) even reported bruchid larvae feeding on developing cotyledons of germinating seeds. So, bruchids will not kill all infested seeds, but their competitive status may be changed.

Several authors (Halevy 1974, Doran *et al.* 1983, Coe and Coe 1987) concluded that bruchid infestation will make the seed coat more permeable to water and make the seeds initiate germination earlier than uninfested seeds. Whether this will give any competitive advantages depends on the conditions, but Ernst *et al.* (1989) reached the conclusion for *A. tortilis* that any advantage of bruchid attack for seed germination will be counteracted by the disadvantages by damaging cotyledons and/or embryo leaving the infested seeds less competitive. Damaged seeds may germinate, but they will probably not survive for long.

4. GERMINATION OF HARD-SEEDED LEGUMES

A general description of the seeds of Sahelian woody legumes will be the basis for understanding dormancy and the natural and artificial pretreatments applied to the hard seeds before germination. The breaking of dormancy, softening of seed coat and water uptake are crucial points in the life cycle of plants, especially those adapted to the semi-arid climate of Sahel. Natural germination will be related to fire, browsers, predation and drought.

Seed characteristics
A seed is the result of fertilization of ovules and the development has been described in detail elsewhere (*e.g.* Bhatnagar and Johri 1972). Although Sahelian woody legumes show some variation in characteristics of the seed, the fundamental characters are the same as for the rest of the legume family. They show some variation in size, shape and weight from the small rounded seeds of *Dicrostachys cinerea* (1000 seeds weigh 26 g) to rather large seeds of *Tamarindus indica* (1000 seeds weigh 400–500 g) (Maydell 1986). Also within species, individuals or even within single pods some variation may occur and this stresses the importance of careful selection of seeds before planting.

Seed coat thickness of various acacias is clearly correlated to their dispersal syndrome: typical wind dispersed seeds have thin seed coats (< 300 µm; *A. senegal, A. ataxacantha, A. macrostachya*), while typically ungulate dispersed seeds have thick seed coats (> 450 µm thick; *A. albida, A. sieberiana, A. nilotica*) (Nongonierma 1978, Schmidt 1988).

A general description of legume seeds can be found in Gunn (1981), but a short description of seed characters of relevance to dormancy and germination will be summarized here (Figures 6 and 7).

Seed dormancy
Dormancy is a condition in a viable seed which prevents it from germinating when supplied with the conditions normally adequate for germination (Willan 1985). It is often associated with a dispersal phase and may be either innate (of genotypes or species), enforced (by cold or drought) or induced (changed dormancy because of changed conditions after ripening)(Harper 1977). Dormancy may also be divided into exogenous (physical impermeability, chemical inhibition or mechanical resistance), morphological endogenous (underdevelopment of embryo) or physiological endogenous (with physiological inhibiting mechanism) or combinations of the three (Willan 1985). In woody legumes of Sahel dormancy is probably mainly physiologically (the impermeable seed coat) controlled by genetical and environmental factors (Rolston 1978).

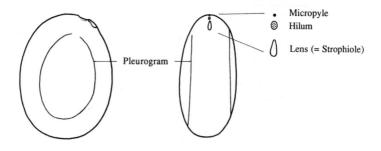

Figure 6. Generalized external morphology of a legume seed. **Pleurogram** (only in some Mimosaceae) — Perhaps functions for drying out mature seeds (Gunn 1981). Develops as a break in the palisade cell layer and it has been shown to be a weak point after boiling (Schmidt 1988). **Micropyle** (small and inconspicuous in Mimosaceae and Caesalpinaceae, sometimes conspicuous in Fabaceae) — the site of passage of the pollen tube between the integuments prior to fertilization, usually a plugged opening in ripe seeds. Always on the opposite side from the lens. **Hilum** — the site of entrance of funiculus, often forming a complex hygroscopic hilar groove in Fabaceae (Hyde 1954). Hilum is a simpler mechanical closure in Caesalpinaceae and some Mimosaceae. Develops as a break in the palisade cells like the pleurogram. **Lens (Strophiole)** — usually conspicuous and dome-shaped in Fabaceae but inconspicuous in Mimosaceae and Caesalpinaceae. An area of weakness, where water often initially penetrates otherwise impermeable testas.

Development of impermeability takes place at 10–12% moisture in *Acacia mearnsii* (Cavanagh 1980) which means when the seeds are ripe. Seeds can be kept soft if frozen, but a proportion of soft seeds (< 10%) are commonly found in seed lots of acacias (Doran *et al.* 1983), indicating that not all seeds develop impermeability, but they seem ready for germination after the first stimulus.

The seed coat dormancy of legumes in semi-arid climate has many important ecological advantages, *i.e.* endozooic dispersal, re-colonization after fire, and escape in terms of time. Perhaps the most important ecological feature of hard-seeded legumes is the differentiated germination of a population under the same stimulus. Seeds with no

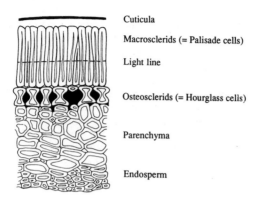

Figure 7. Generalized anatomy of legume seed coat. **Cuticula** — Outer waxy layer covering the whole seed except hilar region. Believed by Schmidt (1988) to hinder water uptake in *Cassia siamea*. **Macrosclerids or palisade cells** (Malphigian layer) — elongated thick-walled cells. A common feature for all hard-seeded legumes, and probably the palisade layer is mainly responsible for impermeability of seed testa (Rolston 1978, Werker 1980/81). **Light line** (only Mimosaceae and Caesalpinaceae) — optical difference due to change in chemical composition of palisade cells. **Osteosclerids or hourglass cells** — believed to be of importance when seeds are maturing and water diffuses from the interior towards the hilum. **Parenchyma** — this layer alone can keep seeds impermeable in some species (Coe and Coe 1987). **Endosperm** — sometimes absent.

pretreatment have been found to germinate in spurts if kept moist over long periods of time (Leistner 1961, Halevy 1974, Nongonierma 1978, Coe and Coe 1987). This polymorphism in seeds needs more research to determine whether it is genetically controlled or the result of different environments (drying rates, predation rates *etc.*). With the unreliable rainfall of the Sahel it would be disastrous if the whole seed crop of a year would germinate after the first isolated showers of the rainy season. Thus, the legumes do not "put all their eggs in one basket."

For the forester, the dormancy of legumes may cause problems. Much effort has been put into hastening germination by removing dormancy of all the seeds in a lot so that they will germinate rapidly and uniformly. The problem has often been described as to raise germination percentages, but in reality it is a question of waiting long enough to reach an ultimate high germination percentage as for instance with *Acacia tortilis* which germinates in spurts over 450 days (Coe and Coe 1987). However, few have been willing to wait 1 1/2 year for germination of seeds in nurseries!

The pretreatment of seeds in the laboratory reflects the way seeds in nature overcome dormancy although the treatments in the nursery often are more drastic. Still, a germination percentage of 100 is seldom reached, but often the ungerminated seeds can germinate if treated once again or in a more severe manner. These extremely hard seeds may be the ones that in nature could survive for 50 years or more and would have a great importance for the ecology and population structure of the species in a disturbed habitat.

Soil seed bank

A bank of viable seeds of hard seeded legumes is believed to exist in the soil awaiting the right conditions for germination. In the literature, the assumption of a viable seed bank in the soil is often found (*e.g.* Coe and Coe 1987), but without data to document it. One of the reasons for this is the time-consuming process of extracting seeds from the soil and much more data on different species in different areas is needed before the regenerative abilities of natural populations under different management strategies are clear.

The only documentation of seed banks found in the literature on Sahelian woody legumes (and native African woody legumes as a whole) is the study on *Acacia sieberiana* in Uganda by Sabiiti and Wein (1987). They found considerable amounts of viable dormant seeds in the soil (800 seeds/m^2 with 2/3 in the upper 2 cm of the soil) under the canopy. However, some information is found on acacias in Australia (Farrel and Ashton 1978, Cavanagh 1980, Grice and Westoby 1987) and introduced acacias in South Africa (Milton and Hall 1981, Pieterse and Cairns 1986). These data cannot be extrapolated to the Sahelian legume species as these species do not have the same predators, the same dispersal mechanisms, *etc.*

Unpublished data by Tybirk, Schmidt, and Hauser give some contradictory information on the soil seed bank of African acacias found in Senegal, Kenya and Zambia, respectively. Schmidt found large numbers of seeds in the soil under *A. seyal, A. hockii,* and *A. mellifera* in Kenya, while Hauser found no viable seeds under the crowns of mature trees of *A. albida* in Zambia. Tybirk found few seeds under *A. nilotica* but no living seeds under *A. tortilis, A. albida, A. senegal* and *A. seyal*. These results indicate that the soil seed bank may exist under certain conditions in certain species, while in other cases no seed bank exists (Tybirk *et al.* in prep.)

This questions the general assumption that hard-seeded legumes have adapted by having a bank of seeds in the soil awaiting the right conditions for germination. In some cases a prolific regeneration has been observed after heavy rains (Schmidt 1988), but often only few or no seedlings emerge (Wickens 1969). Many more systematic investigations are needed before any general statement of this kind can be proven and the implications for management of the Sahel are obvious. If a large seed bank is present, regeneration is much easier, *e.g.* by setting a fire.

Germination in laboratory

The discussion of germination of woody legume seeds will be divided into nursery/laboratory treatment and under natural conditions. The first part is an approximation to what exactly happens when seed dormancy is broken and the seed starts to imbibe and germinate, and the second part describes the factors affecting the process of germination.

Breaking of dormancy. — In the case of Sahelian woody legumes the critical point in breaking dormancy is to make the seed coat permeable to water. As soon as this happens, the seed will germinate rapidly and with high percentages of germination. The following description of the seed morphology and anatomy indicates what could be called the presumed weak points of hard seeds.

In most species the relative thick seed coat cannot be considered a weak point. On the contrary, it is specially adapted to withstand such unfavorable conditions as heat, drought, digestive juice, and mechanical damage. It can be broken by a very severe treatment, but most of the used treatments do not initially make the whole seed coat permeable. After starting to imbibe, the whole seed coat gets soft (Schmidt 1988) and is then probably permeable to water. The "Achilles' Heel" of hard legume seeds is rather the pleurogram, the hilum or the lens (see Figures 6 and 7).

Few authors have tried to discover the weak point of various *Acacia* seeds. Coe and Coe (1987) found that seeds with removed or broken palisade layer may still be impermeable. Schmidt (1988) concluded that permeability is mostly obtained at the hilum or the lens of the seed, but it may be obtained by cracking the palisade layer especially along the pleurogram. The work on Australian acacias (Tran and Cavanagh 1980, Tran 1983, Hanna and Burridge 1983, Hanna 1984) pointed towards the lens being responsible for permeability after various dry or wet heat treatments. Tran (1983) found that the vascular bundles of the mature seeds do not function in water uptake.

Dell (1980) described a "strophiolar plug" in *Albizia lophantha* as responsible for water penetration after wet heating, and water movements are related to vascular bundles of the seed. In some Fabaceae, the hilum is a hygroscopic valve controlling the moisture content of drying seeds and it is therefore suspected to be a weak point in the seed coat (Hyde 1954). The lens has been demonstrated to be the weak area in other Fabaceae (Rolston 1978). Still, many questions about the permeability of hard legume seeds remain unsolved, *e.g.* what happens at the cellular level at the lens or hilum after various treatments and after some hours of imbibition, how does water enter and how is water transported in the seed, and what is the cellular mechanism of softening?

Water imbibition. — When the seed coat is permeable, the seed will take up water if sufficient moisture is available. Chemically or mechanically pretreated seeds have been stored for years before sowing without losing the capability of water imbibition. When seeds are soaked in water, they will imbibe water with varying speed depending on treatment and species. Imbibition is an either/or process, meaning that if water uptake has started, it continues until enough water has been imbibed to complete germination (Schmidt 1988).

When water enters the seed, the seed coat softens, in a physical sense perhaps simply by water uptake of the palisade cells, in the area of water penetration and the softness spreads over the entire seed coat, which may become water permeable by this softening, thus increasing rate of imbibition (Schmidt 1988). Different treatments give different water uptake curves for *Albizia lophantha* (Dell 1980). Boiling for two minutes gives slower imbibition than drilling small holes through the testa and the combined treatments gave the fastest imbibition. After five to ten hours the testa takes up water followed by the embryo resulting in the seeds swelling to the double size within 50 hours after treatment.

Schmidt (1988) found fast imbibition rates of various acacias with average weight increase of 96, 110, and 126 percent of *Acacia hockii*, *A. tortilis* and *A. mellifera*, respectively, within two hours after mechanical scarification. This indeed reflects adaptation to rapid water uptake following short or irregular showers when dormancy has been broken.

Influence of pretreatments in laboratory. — A linear correlation exists between hardness of seed coat of woody legumes and severity of treatment applied to achieve germination (see review by Cavanagh 1980, Doran *et al.* 1983). Unripe (soft) seeds germinate readily after imbibition in cold water while the most dormant old seeds resist imbibition even after very severe treatments. The hardness of seeds increases with age because of suberization of the palisade cells (Cavanagh 1980, Doran *et al.* 1983, Karschon 1975, Kaul and Manohar 1966,

Schmidt 1988) and consequently the pretreatment needed is more severe for older seeds.

Differences within species would probably result from different ecotypes, different genetic composition of the seeds in a sample, different size of seeds, and different position in a pod. A correlation between seed size and thickness of seed coat indicates the correlation with impermeability and dispersal syndrome (Coe and Coe 1987, Nongonierma 1978, see also page 29). Cavanagh (1980) described the larger seeds of acacias as having a more sensitive seed coat remaining soft longer than smaller seeds, but Schmidt (1988) did not find any differences in imbibition of seeds correlated to size or position in the pod. A few selected examples illustrate this:

1. Different dispersal strategies: Stored seeds of *A. senegal* (wind dispersed) need only 3–15 minutes of acid treatment to germinate while stored seeds of the animal dispersed *A. albida, A. nilotica* and *A. tortilis* need 20–120 minutes of acid treatment (Doran *et al.* 1983).

2. Different age of seeds: Fresh seeds of *Cassia sieberiana, Albizia gummifera, Acacia hockii, A. mellifera* and *A. tortilis* had germination of 67-100% after hot water treatment; the same treatment gave germination of 3-31% for old stored seeds (Schmidt 1988).

3. Different severity of treatments: *Acacia nilotica* seeds exposed to different treatments gave the following imbibition percentages after increasing severity of treatment: Untreated, 0%; soaking in alcohol for 24 hours, 0.5%; acid soaking for 15 mins, 1.5%; acid soaking for 30 minutes, 10.5%; dry heating in oven (200° for 3 minutes), 31.5%; hand scarification, 99% (Schmidt 1988).

Table 9 shows a list of various treatments which have been applied to hard legume seeds. The kind and duration of treatment should be considered in each case and depends mainly on species, age of seed lot, and moisture content.

Influence of other factors in nursery. — The seeds of Sahelian woody legumes are probably independent of light/dark conditions for germination (Tewari and Rathore 1973, Mooney *et al.* 1977, Cavanagh 1980, Milton and Hall 1981). Temperature seems to be important in some species mostly with an optimum from 20–30°C; especially higher temperatures can kill seeds (Schmidt 1988, Cavanagh 1980). Most species can germinate in the upper 2 cm soil while some species have been found germinating 10 cm below surface (Cavanagh 1980). But *Acacia senegal* did not germinate below 3.5 cm (Cheema and Quadir 1973) and *A. tortilis* germinates freely down to 4 cm; germinations percentages decreasing deeper in the soil.

Table 9. List of methods for treatment of seeds of Sahelian woody legumes which have been used for obtaining fast and homogenous germination in laboratory or nursery.

Mechanical scarification. Nicking, filing, drilling, soldering iron, shaking in container *etc.* Can be done anywhere on the seed coat. This treatment may be detrimental, but mostly highly efficient, reaching germination percentages close to 100 within few days in most species.

Acid or base. Concentrated H_2SO_4 or NaOH, soaking time variable according to species (*A. senegal*; 3–15 minutes; *A. albida*, 5 minutes; *A. nilotica*, 20–120 minutes; *Cassia siamea*, 10–30 minutes). Often good results obtained with 50–90% germination within 1 day–2 weeks.

Boiling water. Soaking time from few seconds (5 seconds for *A. senegal*) to one hour of boiling for *A. sieberiana*. Germination from 0–90% after treatment (no effect on *Pterocarpus lucens*, 2% germination in *A. nilotica*, 60% in *A. sieberiana*, 90% in *Prosopis juliflora*). May be detrimental in some species, dependent on moisture content.

Hot water. 60–80°C for one to ten minutes will soften coat of some species and rarely be detrimental.

Oven heat. 100–250°C from few seconds to several minutes depending on species and moisture content (*e.g.* 10 minutes at 110°C for *A. hockii*). Acts mainly on the lens and softens the coat in physical sense. May be detrimental.

Temperature fluctuations. Temperature fluctuations with an amplitude of 15°C hastens germination of hard seeded legumes (Quinlivan 1966).

Microwave oven. Dependent on moisture content and species. Cracks the seed coat, but the action is most important on the lens in Australian acacias, no data on African legumes.

Cold water. Soaking for 24 hours. Unripe or soft seeds will germinate immediately.

No pretreatment. Sown directly but kept moist. All species will probably germinate ultimately if kept moist, but wind dispersed *Acacia* species will mostly reach full germination within 4-6 weeks while animal dispersed species will need 15–20 weeks to reach full germination.

Another important factor, when germination in nursery is accounted for, is the amount and frequency of watering in relation to different soil types. Bebawi and Mohamed (1985) made germination trials on six Sudanean acacias with different watering frequencies. All species had highest germination percentages when watered every day and they had clearly reduced germination when the soil was water logged or only watered every third day. *Acacia nilotica* and *A. nubica* did not germinate in water-logged soil but flooding initiated germination of *A. seyal* and *A. tortilis* ssp. *spirocarpa* in nature (Coe and Coe 1987). When watered every third day, *A. seyal* and *A. nilotica* did not germinate, and when watered every sixth day only *A. ehrenbergiana* germinated. These results reflect to some extent the natural habitats of the species, *e.g. A. ehrenbergiana* can germinate under desert conditions, while species often growing in depressions like *A. nilotica* and *A. seyal* need watering every day or constant moist soil without water-logging in the case of *A. nilotica*.

Different soil types also affect germination; seeds in clayey or loamy soil or in animal feces need more heavy and frequent rains for germination than seeds in sand because of higher water retention capacity. Cheema and Quadir (1973) found 90% germination of *A. senegal* in sand and only 16% in sandy clay loam. Leistner (1961) found initially lower germination of *A. erioloba* seeds in manure compared with sand, but ultimately germination was higher in manure. Obeid and Seif el Din (1971) found that more water is needed to reach the same germination percentage for *A. senegal* in clay as in sand; 250 mm rain on sand give higher germination than 500 mm on clay. Coughenour and Detling (1986) found that germination is not affected by nutrient concentration of the media in *A. tortilis*, but water retention capacity of the media is more important. The same conclusion was also reached by Kaul and Manohar (1966) and Cavanagh (1980).

Preece (1971) found that carbon dioxide level influences germination of the Australian *Acacia aneura*; higher levels of carbon dioxide raise the germination percentage in laboratory experiments. The ecological significance of this finding is important if it should be a general tendency in hard-seeded legumes. The level of carbon dioxide in decomposing litter or feces will be higher than in the surroundings, thus hastening the germination of these seeds. However, experiments with African legume species are needed to confirm the suggestion.

Temperature, depth in soil, frequency and amount of water dependent on the water retention capacity of the media are important factors influencing the germination of legume seeds after breaking of dormancy.

Germination in nature
The germination under natural conditions of woody legumes is affected by many factors such as the dispersal strategy, predation, fire, soil, and amount and frequency of rains. Examples of species affected by the different factors will serve to illustrate the complexity of interactions of the mentioned factors on natural germination.

Effects of dispersal. — Dispersal strategy clearly affects the potential sites for deposition of seeds. Wind dispersed seeds may accumulate in rock crevices, rodent burrows, dense bush, *etc.*, while water dispersed seeds mostly end up in alluvial soil in depressions or riverbeds. Animal dispersed seeds are mostly deposited in feces often along animal trails, at watering points, under shade trees, *etc.* These factors give different probabilities of the seed being buried in the soil and different water retention capacity of germination media. However, the differences in nutrient content between feces and soil should not affect the germination, but rather the growth of the seedling.

Different dispersal strategies may also affect breaking of dormancy in different ways. The described methods of breaking dormancy may be applied to conditions met by seeds passing the digestive system of an animal. Chewing and ruminating the pod may give mechanical scarification of seeds similar to treatment in the laboratory perhaps breaking seed coat dormancy. Perhaps bruchid infestation of the seed increases the potential of this "treatment" (Coe and Coe 1987). Digestive juice acts on the seed coat in the same manner as acid treatment in the laboratory and may also break dormancy. The result is faster germination rather than improved germination. The combined effect of chewing, bruchid infestation, digestive juice and water retention capacity of manure gives a complicated picture of factors affecting germination of animal ingested seeds compared to other dispersal strategies.

Wind dispersed seeds, in general, seem to be less dormant or have a larger proportion of soft seeds at the time of dispersal than animal dispersed seeds (Bebawi and Mohamed 1985, Coe and Coe 1987, Schmidt 1988, see page 35). With less severe "treatment" more wind dispersed than animal dispersed seeds will germinate. Perhaps the differences are balanced after dispersal so that animal dispersed seeds need approximately the same stimuli as wind dispersed seeds after dispersal. Species adapted to animal dispersal have higher dormancy and thicker seed coat because they have to be able to withstand passing through an animal and still have some dormant seeds to be dispersed in time.

Effects of predation. — On page 26 some aspects of the influence of seed predation on germination was discussed and when information from this chapter is added, it becomes clear how bruchids can affect germination. A newly hatched larva, which eats its way through the seed coat, inevitably makes this more permeable to water. If it happens in developing green seeds, growth of the testa may close the hole again, whereas if it happens in mature seeds, the effect will be the same as drilling or other mechanical scarification of the coat. The entrance hole of the beetle is small and not readily detectable with the naked eye, and seeds with such holes may constitute part of the seeds classified as soft and germinating without any apparent treatment in experiments.

If development of the larva continues in the seed, it may or may not kill the embryo, but in all cases it weakens the nutritive package of the seed compared to uninfested seeds. The fully developed larva eats its way through the seed coat leaving only a transparent window before pupating (Southgate 1983). The window is only the cuticula and perhaps the outermost part of the palisade cells, and with this natural "treatment" (drilling from the inside), the seed coat must be ready to germinate following no or little pretreatment. However, even seeds with such transparent windows will often be included in seed samples considered undamaged, and if the embryo has not been damaged they will be interpreted as a part of the soft seeds, which they functionally are, in the gemination trial.

The germination capacity of seeds with such transparent windows will be approximately the same as for seeds in which the life cycle of the beetle has been completed leaving the round exit hole. The latter type of seeds will always be considered damaged, and some are capable of germinating (Coe and Coe 1987, Ernst *et al.* 1989, Hauser unpublished report), but will probably not be able to withstand the passage through the digestive tract of an animal, because digestive juice readily enters the seed and digests its content.

Finally, the dormancy of seeds with external damage done by insect larvae, millipedes, rodents, ants, *etc.* (Seif el Din and Obeid 1971a, Hauser unpublished report) will be reduced or removed, and the seeds will be able to germinate without further pretreatment, but they will suffer the same competitive disadvantage as bruchid infested seeds losing part of the nutritive package of the seed.

Effects of fires. — Experiments with dry heating for breaking dormancy (Cavanagh 1980, Schmidt 1988) indicate that the common fires in Sahel have significant influence on germination of woody legumes. This has been observed in nature of Australian acacias (Everist cited in Preece 1971, Milton and Hall 1981, Pieterse and Cairns 1986) and for two species of African acacias; *A. sieberiana* in Uganda (Sabiiti and Wein 1987) and *A. hockii* in Kenya (Schmidt 1988).

Permeability of *A. hockii* seeds in the upper 0.5 cm of the soil is improved after a fire, but there is no significant improvement of seeds from deeper soil layers (Schmidt 1988). Sabiiti and Wein (1987) found increased germination of seeds in the upper two cm of the soil while deeper laying seeds were unaffected by the fire intensities used in the field experiment. In both cases it is probably only hastened and not improved germination as stated by Sabiiti and Wein (1987). Fires will, however, often kill many seeds on the surface but seeds in deeper soil layers will get the positive effect on permeability, so any fire will break dormancy and initiate germination of some seeds in the presumed seed bank (see page 32, Figure 8A).

Selective killing of developing bruchid larvae before the embryo of the seeds is damaged (discussed on page 25) may be a more important ecological effect of fire. It prevents a part of the seeds from being consumed and in the long run that is more important than the hastened germination.

Figure 8A. The frequent fires in Sahel severely influence regeneration of woody legumes. Photo: Assane Goudiaby. **8B.** Nursery plants are vulnerable during the first years after transplantation.

Effects of soil and water. — Water balance for germinating seeds is directly related to soil conditions (page 37). Different dispersal strategies result in deposition of seeds in different soil types and dispersal in time may result in a different precipitation pattern for germination. Over longer time periods species may colonize new areas if rainfall patterns change and the soil requirements suit the species under the changed water balance.

Differences in water balance of different soil media, *e.g.* sand and clay, give different germination percentages of woody legumes. Such differences also influence the general distribution of many species in the northern and more dry parts of Sahel where they are found on sandy soils rather than on clay because the water retention capacity is much lower in sand than in clay (Maydell 1986).

The acidity of soil types influences breaking of dormancy; permanent action of weak acids in the soil soften the seed coat in the same way as short laboratory treatments in strong acids (Nongonierma 1978).

5. SEEDLING GROWTH

Establishment of the seedling and growth of young plants is another critical phase in the life of Sahelian woody legumes. This section describes development of the seedling, vegetative regeneration potentials and influence of rainfall, soil types, predating animals, fire, light, *etc.* on growth of the young plants.

Development of the seedling

Germination in most of the Sahelian woody legumes is phaneroepigeal with foliar cotyledons, which is the predominant type in open tropical habitats (Duke and Polhill 1981). Subsequent development of leaves and roots depends on the species, and the speed of growth depends on several factors.

Root development is critical and relatively fast in dry zone legumes. The roots of *Acacia senegal, A. albida,* and *A. tortilis* reached a depth of > 100 cm in three months (Gupta *et al.* 1973) and *A. albida* developed a root length of 670 cm in six months (Anonymous 1988). The development of lateral roots seems also to depend on species and different "treatments." In *A. senegal* lateral roots will normally only develop below 25 cm, but simulated browsing may either suppress when browsing during the first two weeks of development, or stimulate when older seedlings are browsed, development of lateral roots (Seif el Din and Obeid 1971b).

Growth of above ground parts of the plants also depends on species, ecotypes, and conditions. The speed of growth is important when selecting species for afforestation projects, and indeed it has great influence on the regeneration and the duration of protection of new plantations. Several tree species have been introduced to Africa because of their fast growth, *e.g. Prosopis juliflora, Cassia siamea,* and *Leucaena leucocephala.* This does not necessarily mean that these species are fast growing in Sahel where relatively unfavorable conditions prevail.

The mean annual height increment of healthy plants varies from 14 cm in *Albizia lebbeck* to 66 cm in *Prosopis juliflora,* but other examples of height increment may show more variation: *Cassia siamea* can reach a height of five m in two years, *A. senegal* may be only 10 cm when eight months old but may reach 90 cm in three years, *A. albida* has been found to vary from two m in five years to 5.5 m in three years depending on the location (Wickens 1969, Dyson and Thogo 1976, Muthana and Arora 1980). Whether the plants have been sown directly or transplanted from nurseries is also important. The start in a nursery can give the plant a good offset, but transplantation may be the most dangerous part of its lifetime (Figure 8B). Mortality is often high during the first dry season after transplanting of nursery plants (Gosseye 1980).

Vegetative regeneration

Asexual regeneration is rarely found in Sahelian woody legumes in nature. The only species reported in the literature to have natural clonal growth is *A. albida* growing in Israel, where one tree during 50–60 years developed into a clone covering 1500 m^2 by root suckers (Karschon 1976). Also *Acacia albida* grown in India developed root suckers (Ahmed 1987). However, when human management is considered, it may often be useful to propagate trees asexually to obtain homogeneous genetic material as well as fast establishment of young plants.

Several methods have been developed for vegetatively multiplying individual trees: shoot or branch cuttings, root cuttings, and meristem culture. In the case of dry zone legume trees these methods are almost exclusively used in nurseries and laboratories. The most common utilization of woody legumes by local people in the field is by pollarding/ pruning/ looping of the trees. This method is often used for producing fuelwood and fodder for cattle and several species such as *Acacia albida* and *Pterocarpus lucens* are suitable. In general only a few species, for example *Entada africana*, will sprout from stumps after fire (Brookman-Amissah *et al.* 1980), but more species tolerate looping of the crown and this method will give the best crop every 1–2 years.

Shoot, branch, or root cuttings for propagation are in principle the same. A suitable piece (20x2–3 cm) is cut of a plant and planted in a pot, watered, and a new plant may develop. In some cases hormones are applied to stimulate root development and after some months the seedling may be transplanted in the field. The technique is useful in some species, but Gosseye (1980) and Piot (1980) find the method difficult to apply in the field in places where the annual precipitation is below 500 mm. The meristem culture method is just about being developed for scientific purposes (Gassama 1989) in some *Acacia* species, but is still quite far from being applicable as a large scale plantation technique.

Factors influencing growth

Whether regeneration is natural or artificial, sexual or vegetative, the young plants are sensitive to a variety of factors affecting survival and growth. The most important of these will be described here.

Water. — Amount and distribution of water is the most important factor determining survival and growth of young plants and is closely correlated to soil type and light exposure. The seasonal climate of Sahel makes sowing or transplanting trees in the beginning of the dry season doomed to failure. When planters rely on natural rainfall, seeds must be sown late in the dry season and seedlings transplanted in the beginning of the rains. Surviving the first rainy season is essential, and years with a month without rain in the middle of the rainy season is detrimental to seedlings, both directly sown and transplanted.

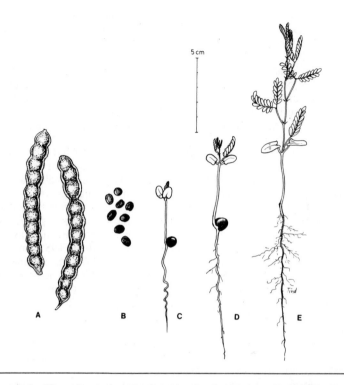

Figure 9. Seedlings of several acacias (here *Acacia nilotica*) appear in the beginning of the rainy season, but very few will survive the first year.

Growth of shoots and roots of several acacias is optimal when the seedlings are watered dayly, and growth is reduced by 62% and 84% when watered only every second and forth day, respectively (Bebawi and Mohammed 1982). *Acacia seyal* is very sensitive to water supply, *A. albida*, *A. nilotica*, and *A. ehrenbergiana* are less sensitive while *A. tortilis* ssp. *spirocarpa* is intermediate. Cheema and Quadir (1973) found optimal growth in *A. senegal* when watered only every seventh day due to low aeration of soils when watered more often.

Schmidt (1988) studied natural regeneration of several acacias after 25 mm of rain in Kenya. *Acacia mellifera* seedlings appeared from five to nine days after the rain in 57 seedlings on 18.84 m^2, while ten *A. nilotica* seedlings appeared on the same area from seven to eleven days after the rain. So the sole stimulus of rain gave prolific regeneration of several species, but three weeks later only very few or no seedlings had survived. The same has been observed by the author with seedlings of *A. seyal, A. tortilis*, and *A. nilotica* in Senegal where after four weeks of drought in July/August 1989, apparently no seedlings from this year had survived (Figure 9).

The most favorable sites in such years are depressions or temporary pools where drying out of the soil is slower leaving the seedlings time to develop the root system and to reach permanent water table. Experiments are going on in Senegal on the germination and growth of *A. senegal* in relation to rainfall (Danthu, pers. comm.).

Soil type. — Water retention capacity of soil, correlated with nutrient status and salt or sodium levels, is not only important for germination (page 37), but also for growth of seedlings. Only a few studies have been conducted to find the optimal conditions for seedling growth. Cheema and Quadir (1973) found that for *A. senegal* coarse soil texture with medium water retention capacity (+/÷ organic matter) and pH 7.5 were optimal. Soil with a mixture of sediment and manure increases the growth rate of *A. tortilis* seedlings (Gupta and Muthana 1985) and mean annual height increment was 44 cm on rocky and 61 cm on sandy soil (Muthana and Arora 1980). In trials with *A. albida* in India the best growth was obtained in sandy loam with farm manure (Ahmed 1987) and for many of the animal dispersed species similar conditions may be the conditions for growth in nature. The high ammonium content of animal feces, however, may kill seedlings after germination (Wickens 1969, Janzen 1971, Pijl 1982) so the feces may be detrimental. Positive effects may be found when secondary dispersal (page 7–8) has occurred or after some decomposition and leaching of the feces.

Phosphorus has been mentioned as a limiting factor to growth in seedlings of *A. nilotica* and inoculation of VA-mycorrhizal fungi increased root and shoot growth of *Leucaena leucocephala* (Michelsen 1989). Mycorrhiza inoculation is believed to have great impact on phosphorous uptake (Högberg 1989, Michelsen and Rosendahl 1990), but also *Rhizobium* inoculation in early growth phase has been shown to double the growth of *A. albida* seedlings (Anonymous 1988). Most Sahelian woody legumes fix nitrogen by *Rhizobium*, but often trapping or growth of bacteria fails in nature due to drought or pH below 5–5.5 and above 8.5–9 (Habish 1970, Allen and Allen 1981). In the nursery nodulation often occurs but after transplanting nodulation seems closely related to rainfall (Felker 1978, Bakolimalala 1989).

Salinity and alkalinity of soil may be a determining factor in semi-arid areas. *Parkinsonia aculeata, Prosopis juliflora, Leucaena leucocephala*, and *Acacia nilotica* are said to tolerate some salinity, while *Albizia lebbeck, Cassia siamea,* and *A. tortilis* tolerate only low salt and sodium concentrations (Tomar and Yadav cited in Fagg and Greaves 1990b, Tomar and Gupta 1985).

Predation on seedlings. — Young plants may be subject to predation by a variety of animals and fungi. They are nutritious, palatable, and contain sufficient water for many animals. The emergence of seedlings in the beginning of the rainy season coincides with emergence of offspring of insects and grasshoppers, but also termites, millipedes, spiders *etc.* are known to feed on the succulent seedlings (Giffard 1966, Seif el Din and Obeid 1971a). Removal of cotyledons of three day old seedlings, simulating insect damage, reduced shoot and root growth and raised mortality of several Sahelian acacias (Rahman and Dafei cited in Fagg and Greaves 1990b).

Impact of mammalian herbivores has been mentioned by several authors (Giffard 1966, Khan 1970, Obeid and Seif el Din 1970, Cheema and Quadir 1973, Gosseye 1980, Schmidt 1988) and Seif el Din and Obeid (1971a,b) made detailed studies of simulated browsing on *A. senegal*. They found that seedling survival is best when seedlings are browsed during the first two weeks, due to compensatory effect of cotyledons, or after eight weeks of age when the plant has established its roots. However, in no cases were more than 60% of the seedlings killed, so in general seedlings will often survive one browsing. Repeated browsing will exhaust the young plant and eventually kill it or transform the plant to cushion-like growth (Harker 1959, Cheema and Quadir 1973, Piot 1980).

Most species are heavily browsed even though the strong unpleasant smell of crushed seedlings of many acacias may repel ungulates (Schmidt 1988). The smell persists until the seedlings starts to lignify and develop spines. *Prosopis juliflora* seems to be the only species which is not browsed heavily and is thus the only one not needing protection against cattle (De Candolle 1962, Pedersen 1980). It is mostly the small ruminants such as goats and sheep that browse on developing plants (Piot 1980, Harvey 1981), and seedlings need protection for the first two to three years until they are out of reach of goats (Giffard 1964, 1966, Piot 1980, Ahmed El Houri 1986, Anonymous 1988). Regeneration is usually out of the question if no protection is possible.

Fire. — The common practice in Sahel of management of grazing areas by annual burning has severe influence on the growth of legume seedlings. Contradictory information on the impact of fire is found in the literature. It has been argued that fire promotes acacias (Coe and Coe 1987) as well as that fewer fires will result in survival of more seedlings of *A. sieberiana* (Sabiiti and Wein 1987), and there is no doubt that the influence is complex depending on the species and the time, intensity and frequency of fires as well as other management strategies in combination with fires (Tolsma 1989).

Sprouting of seedlings after fire is a well known phenomenon (Sabiiti and Wein 1987, see page 39), but survival of seedlings certainly depends on the time until the next fire. Very young seedlings may simply be killed by fires, but when seedlings are some months old, they will often be able to survive a fire. Established young plants may survive repeated burnings because the root will survive and the plants are usually burned in the dry season, when the plants are in the resting phase. *Acacia albida* is known to develop a thick and deep root if browsed repeatedly (Giffard 1964) and this is probably also the case when burned repeatedly. Thus, the root needs protection to develop into a tree.

In studies of woody vegetation in more humid African savannas it has been found that fire in April gives lower tree density, *i.e.* more seedlings are killed, than fire in November or no fire at all (Brookman-Amissah *et al.* 1980). This is probably because the fire becomes more intense (drier fuel load) late in the dry season. The influence of fire on regeneration of woody legumes in Sahelian vegetation is not well known, but probably they are best adapted to low intense fires which are common in dry habitats. However, fires are frequent in the Sahelian zone in Senegal (Langaas, pers. comm.) and occur shortly after the rains in October–December, when high fuel load is present.

When discussing tree planting in Sahel in general, fire protection is needed to ensure fast growth of the young plants to escape both fires and browsing. Two to four years of protection is recommended for woody legumes (Anonymous 1960, De Candolle 1962, Gosseye 1980).

Competition. — Establishment of seedlings of *A. tortilis, A. nilotica*, and *Dicrostachys cinerea* depends on the amount of light (Khan 1970, Smith and Goodman 1986, Smith and Shackleton 1988, Tolsma 1989). Smith and co-workers found that more seedlings occur below the crown of parent trees, but fewer seedlings establish and develop. Soil moisture and nutrient content of the soils were sufficient for establishment, but light was the limiting factor under the crowns. The biomass of shoot and root is reduced and this reduces the chances for survival during the dry season; the woody legumes in Sahel are indeed light lovers.

Knoop (cited in Smith and Shackleton 1988) found that there was a significant increase in establishment of seedlings of *A. tortilis* and *A. nilotica* in South Africa when the grass cover was removed. This may be due to light competition, but competition for water may also be the case. The shallow roots of grasses compete for the water in the top 50 cm of soil with the establishing legume seedling. Soil water depletion is faster under the crown of trees (Smith and Shackleton 1988) due to high transpiration rate of the adult tree and this may also be part of the reason for the lack of establishment of legume seedlings under crowns. Giffard (1966) and Hauser (unpublished report) did not find any regeneration of *A. albida* under the crowns of adult trees, but this may be due to a combination of several factors (light, soil water, trampling of animals seeking shade *etc.*).

When a seedling has established, the competition for soil nutrients becomes important as well. Growth rates have been shown to be higher when manure is mixed with the soil (page 46) and on poor soils competition with other plants for soil nutrients may become important for growth and development.

6. IMPLICATIONS FOR MANAGEMENT

The first chapters of this book have summarized existing knowledge concerning the main bottlenecks in the regeneration of Sahelian woody legumes. If the risk of predation for seeds, risk of unfavorable germination in time or space, risk to be eaten, burnt, or killed by drought as young plant, *etc.* are summed up, the chances to survive and develop into a mature tree seems next to zero. Still, there must be a chance because young trees are indeed found in some areas.

It has to be emphasized here that it is not desireable that all the included legumes should spread in large areas of Sahel. This goes for some of the thicket forming species such as *Acacia macrostachya, A. ataxacantha, Dichrostachys cinerea*, and *Mimosa pigra*. Bush encroachment is a problem in some areas where these thorny shrubs form impenetrable thickets making the areas useless (Tolsma 1989).

In other areas the problem is the opposite: woody vegetation disappears and no natural regeneration is possible due to the increased pressure on agricultural and grazing land and subsequently changed management. In many cases increase in awareness and information concerning the bottlenecks mentioned in this book are essential in order to develop slightly different ways of utilizing the areas, which in the long run may be much more efficient for regeneration of trees within their natural distribution areas.

It may be worth while to summarize the main problems and suggest improvements for each case. For the different dispersal strategies, not much can be done for improving wind and water dispersal, apart from collecting and dispersing seeds by hand in favorable sites at favorable time for germination. For animal dispersed species it is obvious that collecting pods as fodder for domesticated animals at the right time and letting the animals defecate in favorable sites such as depressions or fenced areas in the beginning of the rainy season can be advantageous (Tybirk 1991).

The best ways of reducing impact of bruchids on seed crops would be to collect seeds as early as possible after full development and store seeds or pods in airtight containers or in carbon dioxide to kill developing larvae. Traditional methods such as control with kernels of *Azadirachta indica*, tobacco plants, chilipepper, black pepper, *etc.* and extracts of *A. indica*, garlic, citrus oils, palm kernel oils, coconut oil, mustard, tamarind ashes, salt, sand, sawdust, smoke, *etc.* have also been recommended for bruchid control (Johnson 1983).

Looking further ahead, basal investigations on the interactions between bruchids and their parasites or predators (wasps, mites, *etc.*) may give valuable information on the possibilities of biological control, but at the moment the potential of this method cannot be estimated.

Concerning timing of germination, the best solution is probably to expose a seed lot to as many different treatments and conditions as possible with some seeds in acid, some seeds fire exposed, some seeds scarified and sown in as different habitats, soils, depths, cultivation treatments, *etc.* as possible. This will give the genetic varieties different possibilities to germinate and establish. This would reflect the differences in stimuli that natural populations are exposed to during many years. If working on a small scale, the best treatment to reach full germination is to scarify the seeds, but this can be costly. When making priorities between money and germination efficiencies, the listed treatments combined with the species list can be valuable.

Trees need protection against fire and browsing during the first two to four years after germination. This is difficult for obvious social reasons such as land tenure, user rights, traditions, *etc.*, but perhaps protection is possible in small areas. However, in any case the survival rate of the seedlings is very low (Tolsma 1989).

The most important general conclusions about regeneration of Sahelian woody legumes are that even though there are many odds in nature against natural regeneration, still it does take place. Even though the last two decades have brought several dry periods in Sahel and the term desertification is found everywhere in papers from donor agencies involved in the area, the resilience of vegetation in areas with dynamic vegetation well adapted to fluctuating conditions has to be kept in mind. The simultaneous occurrence of a year of good seed crop, low seed predation, and good rains is rare, but is the prerequisite of a good rejuvenation of the woody legumes in Sahel.

This book has not led to any obvious conclusion to improve regeneration of woody legumes in Sahel in general, but shows the complexity of interactions between the different factors affecting regeneration. Hopefully, the book can help to take the described bottlenecks into consideration in the particular cases where restoring of woody vegetation is desirable.

7. SPECIES SUMMARIES

Further information about the listed species with descriptions, distributions, site requirements and uses can be found in Maydell (1986)

Acacia albida Del. [= *Faidherbia albida* (Del.) A. Chev.]
(MIMOSACEAE)
Dispersal. — The indehiscent pods are dispersed by ungulates and most likely also by water.
Seeds characteristics. — 10–25 seeds/pod. Seed weight 38–210 g/1000, rounded seeds.
Seed pests. — 38% of all stored seeds infested. One larva of *Bruchidius* sp. attacks 2–3 seeds of each pod. Percentages in parenthesis are examples of known infestation rates for the individual species: *Bruchus natalensis, Bruchidius auratopubens* Decelle(10–82%), *B. aurivillii* (41%), *B. cadei* Decelle (4–86%), *B. platypennis* Decelle (75%), *B. pygidiopictus* Decelle (10–80%), *B. silaceous, B.* sp. near *rufulus, Bruchidius sp., Caryedon excavatus* Decelle (0–81%), *Pachymerous longus, P. pallidus, Tuberculobruchus pygidiopictus* Decelle.
Germination. — To obtain fast and high germination percentages various pretreatments have been applied, *e.g.* hot water or boiling few seconds, acid treatments from 5–60 minutes, mechanical scarification. These treatments give germination of 80–100% in 2 weeks, while untreated seeds reach about 70% germination in 20 weeks. Soft seeds will germinate readily without pretreatment. Direct seeding is possible, but protection is then necessary
Vegetative propagation and seedling growth. — Can be quite effective by root suckers and branch cuttings transplanted in the beginning of rainy season where precipitation exceeds 500 mm/year. Coppicing abilities well known in crown and from stumps. Clonal growth has been observed and in vitro culture has been developed. Root and shoot length of seedlings are more or less equal, up to >100 cm in 3 months (5 m in less than 1 year has been observed), shoot growth of 115 cm/year on sandy loam with manure. Severe predation of seedlings which need full sunlight, but may resprout after some browsing and burning. Growth may be twice as fast when *Rhizobium* inoculated in nursery.
References. — Giffard 1964, 1966, 1971; Prevett 1967; Radwanski and Wickens 1967; Wickens 1969; Gupta *et al.* 1973; Karschon 1976; Felker 1978; Nongonierma 1978; Gosseye 1980; Piot 1980; Allen and Allen 1981; Doran *et al.* 1983; Tonder 1985; Maydell 1986; Ahmed 1987; Anonymous 1988; Hauser unpublished report; Sniezko and Stewart 1989.

54 Knud Tybirk

Acacia ataxacantha DC. (MIMOSACEAE)

Dispersal. — The dehiscent pod is dispersed by wind and probably by ungulates in some cases.

Seed characteristics. — Flattened seeds (average 4.5/pod; 40–90 g/1000 seeds).

Seed pests. — 20% of all seeds in stores are attacked. Percentages in parenthesis are examples of known infestation rates for the individual species: *Bruchidius cadenati* (Pic) (5–18%), *B. silaceus* Fahr. (10–18%), *B. submaculatus* Fahr. (9%), *Bruchidius* sp., *Caryedon mauritanicus* Decelle (10–55%), *C. pallidus* (Ol.), *Tuberculobruchus* sp.

Germination. — Untreated seeds reach 44% germination in 4 weeks, while scarified seeds reach 72% germination in the same period.

Vegetative propagation and seedling growth — No data. Nitrogen fixing.

References. — Buchwald 1895; Prevett 1967; Nongonierma 1978; Allen and Allen 1981; Maydell 1986.

Acacia dudgeoni Craib ex Holl. (MIMOSACEAE)

Dispersal. — The dehiscent pod is dispersed by wind and probably by ungulates in some cases.

Seed characteristics. — Flattened seeds (average 3.9/pod), 95–125 g/1000 seeds.

Seed pests. — 7% of all seeds in stores are attacked. Percentages in parenthesis are examples of known infestation rates for the individual species: *Bruchidius catenati* (Pic) (11–18%), *B. silaceus* Fahr. (4%), *B. tougourensis* Decelle (2%), *Caryedon mauritanicus* Decelle (12%).

Germination. — 22.5% of untreated seeds germinated in 6 weeks, while 65% of scarified seeds germinated in 2 weeks.

Vegetative propagation and seedling growth. — No data. Possibly nitrogen fixing.

References. — Brookman-Amissah *et al.* 1980; Allen and Allen 1981; Varaigne-Labeyrie and Labeyrie 1981; Maydell 1986.

Acacia ehrenbergiana Hayne (MIMOSACEAE)

Dispersal. — Dehiscent pod (similar to *A. seyal*) with hanging seeds dispersed by wind, ungulates and possibly birds.

Seed characteristics. — Flattened small seeds (average 7.1/pod), 24 g/1000 seeds.

Seed pests. — 97% of all seeds have been reported infested in stores. Percentages in parenthesis are examples of known infestation rates for the individual species: *Bruchidius sahelicus* Decelle (65–100%), *B. uberatus* Fahr. (86%), *Caryedon sahelicus* Decelle (65–100%).

Germination. — 32% of untreated seeds germinated in 5 weeks, while only 10% scarified seeds germinated in 2 weeks. The species has been considered a medium dormancy species, and it may germinate when watered only every 6th day.

Vegetative propagation and seedling growth. — Removing cotyledons of seedling reduced root and shoot production and increased mortality. Nitrogen fixing.

References. — Nongonierma 1978; Allen and Allen 1981; Bebawi and Mohammed 1982, 1985; Rahman and Dafei cited in Fagg and Greaves 1990b; Maydell 1986.

Acacia gourmaensis A. Chev. (MIMOSACEAE)

Dispersal. — The dehiscent pod is dispersed by wind and sometimes by ungulates.

Seed characteristics. — Average 2.5 flattened seeds/pod, 55–90 g/1000 seeds.

Seed pests. — 24% of all seeds have been reported infested in stores. Percentages in parenthesis are examples of known infestation rates for the individual species: *Bruchidius cadenati* (Pic) (14%), *B. submaculatus* Fahr. (14%), *Caryedon mauritanicus* Decelle (14–32%).

Germination. — 62% of untreated seeds germinated in 5 weeks, 36 % scarified germinated in 1 week.

Vegetative propagation and seedling growth. — Seedling survival is better after late fires than early fires.

References. — Nongonierma 1978; Brookman-Amissah *et al.* 1980; Allen and Allen 1981; Maydell 1986.

Acacia laeta R. Br. ex Benth. (MIMOSACEAE)

Dispersal. — The dehiscent pod is dispersed primarily by wind, possibly also by ungulates.

Seed characteristics. — Average 2.2 flat seeds/pod, 83–106 g/1000 seeds.

Seed pests. — 32% of all seeds have been found predated in stores. *Bruchidius cadenati* (Pic) was responsible for 27%.

Germination. — 2% of stored untreated seeds germinated in 3 weeks, while 71.5% of scarified seeds germinated in 1 week. Boiling has also been recommended.

Vegetative propagation and seedling growth. — No data.

References. — Buchwald 1895; Nongonierma 1978; Allen and Allen 1981; Depierre and Gillet cited in Fagg and Greaves 1990b; Maydell 1986.

Acacia macrostachya Reichenb. ex Benth. (MIMOSACEAE)

Dispersal. — The dehiscent pod is primarily dispersed by wind but possibly also by ungulates.

Seed characteristics. — Flattened seeds (average 3.7/pod), 66–77 g/1000 seeds.

Seed pests. — 41% of all stored seeds have been found infested. Sometimes one or two species may develop more than one adult beetle /seed. Percentages in parenthesis are examples of known infestation rates for the individual species: *Bruchidius auratopubens* Decelle (49%), *B.*

aurivillii Blanc (53%), *Bruchidius cadenati* (Pic)(24–91%), *B. pennatae* Decelle (17–18%), *B. silaceus* Fahr.(6–91%), *B. submaculatus* Fahr. (13–91%), *B. tougourensis*, *Caryedon mauritanicus* Decelle (48–91%), *Tuberculobruchus natalensis* (Pic) (91%).
Germination. — 5% of untreated seeds germinated in 3 weeks, while 77% of scarified seeds germinated in 1 week.
Vegetative propagation and seedling growth. — No data.
References. — Nongonierma 1978; Allen and Allen 1981; Varaigne-Labeyrie and Labeyrie 1981; Maydell 1986.

Acacia macrothyrsa Harms (MIMOSACEAE)

Dispersal. — The dehiscent pod is dispersed by wind and possibly also by ungulates.
Seed characteristics. — Flat seeds (average 8.3/pod), 65–112 g/1000 seeds with thin parenchyma layer (180–275 μm).
Seed pests. — 6% of all seeds stored have been found infested, in some trials *Bruchidius senegalensis* alone has infested 34% of the seeds.
Germination. — Untreated seeds had 13% germination in 7 weeks, while 70.5% of scarified seeds germinated in 1 week.
Vegetative propagation and seedling growth. — No data, but nitrogen fixing.
References. — Nongonierma 1978; Allen and Allen 1981; Maydell 1986.

Acacia mellifera (Vahl) Benth. (MIMOSACEAE)

Dispersal. — The dehiscent pod is dispersed by wind and ungulates.
Seed characteristics. — Flat seeds.
Seed pests. — *Bruchidius sp.*, *Caryedon pallidus* (Ol.).
Germination. — After hot water treatment of seeds, 28–48% germinated within 2–4 weeks. 100% of seeds from goat dung imbibed readily. Seeds increase 126% in weight 2 hours after imbibition has started.
Vegetative propagation and seedling growth. — Removing cotyledons of seedlings increased mortality and reduced growth. Seedlings appear within 9 days after rain, but often none will survive. Nitrogen fixing.
References. — Buchwald 1895; Nongonierma 1978; Allen and Allen 1981; Rahman and Dafei cited in Fagg and Greaves 1990b; Tonder 1985; Maydell 1986; Schmidt 1988.

Acacia nilotica Willd. ex Del. (MIMOSACEAE)

Dispersal. — The indehiscent pod (sometimes segmenting) is dispersed primarily by ungulates, but also water may be important in some populations and wind has been mentioned.
Seed characteristics. — Large rounded very hard seeds with a thick parenchyma layer (665–725 μm) (numbers of seeds /pod varies from 4.6–10.8), 133–200 g/1000 seeds.
Seed pests. — 36–39% of stored seeds have been reported infested. Percentages in parenthesis are examples of known infestation rates for the

individual species: *Bruchus maculatus, B. silaceus* (Fahr.), *Bruchidius analis, B. albonotatus, B. albosparsus, B. bedfordi, B. baudoni* (Caill.), *B. centromaculatus, B. luteopygus* (Pic) (2%), *B. quadrimaculatus, B. senegalensis* (17–84%), *B. sieberianae* Decelle (13%), *B. submaculatus, B. theobromae, B. uberatus* (Fahr.) (0–84%), *Bruchidius sp., Caryedon capicola* (Motschoulsky) (28–84%), *C. albonotatus* (Pic), *C. interstinctus* (Fahr.) (4%), *Pachymerus cassiae, P. longus, P. pallidus, Tuberculubruchus natalensis* (Pic) (26%). Also Curculionidae and Cerambycidae (*Enaretta brevicaudata, E. castelnaudi, E. paulinoi)* have been reported on seeds.

Germination. — This species has been considered a high dormancy species that does not germinate without pretreatment. Fresh seeds need no pretreatment, while old non-treated seeds gave 19–63% germination in 15 weeks, old scarified seeds gave 65–92% germination in 1 week depending on the subspecies. Also hot water, boiling, oven heating, and acid (60–120 minutes) treatments have been applied to hasten germination. Seeds from cattle dung gave 33–51% germination, while seeds from sheep dung gave 15% germination. No germination when waterlogged but still germination needs watering every third day.

Vegetative propagation and seedling growth. — Cuttings treated with hormones will root. Contradictory information exists on coppicing abilities. Seedlings will tolerate moderate salt and sodium concentrations, but seedling growth is limited by water and phosphorus. Nitrogen fixing. Mychorrizal inoculation increases water and nutrient uptake. Seedlings suffer from shading, browsing (sheep eat more seedlings than cattle), cracking of soil, competition with grasses, *etc.* Seedling growth has been reported from 46–100 cm/ year.

References. — Buchwald 1895; Lamprey 1967; Prevett 1967; Khan 1970; Nongonierma 1978; Muthana and Arora 1980; Allen and Allen 1981; Harvey 1981; Varaigne-Labeyrie and Labeyrie 1981; Bebawi and Mohammed 1982; Doran *et al.* 1983; Khan cited in Fagg and Greaves 1990a; Singh cited in Fagg and Greaves 1990a; Tomar and Yadav cited in Fagg and Greaves 1990b; Mathur *et al.* 1984; Tomar and Gupta 1985; Tonder 1985; Maydell 1986; Schmidt 1988; Smith and Goodman 1986; Knoop cited in Smith and Shackleton 1988; Michelsen 1989; Michelsen and Rosendahl 1990; Tolsma 1989; Tybirk 1989.

Acacia pennata (L.) Willd. (MIMOSACEAE)
Dispersal. — The dehiscent pod is dispersed by wind and possibly also by ungulates.
Seed characteristics. — Seeds are compressed, average 9.1/pod, 107–111 g /1000 seeds.
Seed pests. — 13% of stored seeds have been reported infested. Percentages in parenthesis are examples of known infestation rates for the individual species: *Bruchidius cadenati* (Pic) (8–18%), *B. pennatae* Decelle (4–36%), *Caryedon mauritanicus* Decelle (12%).

Germination. — 19% untreated seeds germinated in 6 weeks, while 75% of scarified seeds germinated in 2 weeks.
Vegetative propagation and seedling growth. — No data.
References. — Buchwald 1895; Nongonierma 1978; Allen and Allen 1981; Varaigne-Labeyrie and Labeyrie 1981; Maydell 1986.

Acacia polyacantha Willd. (MIMOSACEAE)

Dispersal. — The dehiscent pod is dispersed by wind and possibly also by ungulates.
Seed characteristics. — Flattened seeds, average 2.5–6.3/pod, 78–90 g/1000 seeds, parenchyma layer 150–200µm.
Seed pests. — 13% of stored seeds have been found infested. Percentages in parenthesis are examples of known infestation rates for the individual species: *Bruchidius cadenati* (Pic) (19%), *B. campylacanthae* Decelle (7–19%), *B. senegalensis* (Pic), *B. silaceus* (Fahr.) (4–7%), *Caryedon mauritanicus* Decelle (7–19%).
Germination. — 6% of untreated seeds germinated in 6 weeks, while 74% of scarified seeds germinated in 1 week.
Vegetative propagation and seedling growth. — Nitrogen fixing, but otherwise no data.
References. — Nongonierma 1978; Allen and Allen 1981; Maydell 1986; Tybirk unpublished.

Acacia senegal (L.) Willd. (MIMOSACEAE)

Dispersal. — The dehiscent pod is dispersed by wind and also by ungulates.
Seed characteristics. — Flat, rather small seeds, average 3–4.5/pod, 55–83 g /1000 seeds.
Seed pests. — 15–38% of stored seeds have been reported infested. In nature up to 2/3 of the seeds have been found damaged by beetles, rodents and millipedes before germination. Percentages in parenthesis are examples of known infestation rates for the individual species: *Bruchus centromaculatus*, *B. subuniformis*, *B. silaceus*, *Bruchidius albosparsus*, *B. cadenati* (Pic) (3–51%), *B. pennatae* Decelle (6%), *B. petechialis* (Gyllenhal), *B. sahelicus* Decelle (27–77%), *B. senegalensis* (Pic) (25%), *B. submaculatus* (Fahr.) (6%), *B. uberatus* (Fahr.) (3–29%), *Caryedon mauritanicus* Decelle(13–77%), *C. longispinosa* Decelle (19%), *C. pallidus*, *C. sahelicus* Decelle (29–98%), *Pachymerus pallidus*, *Tuberculubruchus natalensis* (Pic) (7%).
Germination. — Fresh green seeds picked off the trees show two types of germination response. Some imbibe readily and germinate while others have slower germination. 47–83% untreated seeds germinated in 3 weeks, while 63–78% germinated when scarified, so in general it germinates easily without severe pretreatment. Germination is indifferent to light /darkness but is best in the upper 2 cm soil and depends on

amount and frequency of showers in sand. More water is needed for germination in clay.

Vegetative propagation and seedling growth. — Shoot cuttings are possible. Seedlings rather slow growing with annual height increment of 30–40 cm, but roots may reach >100 cm depth in 3 months. Growth best when watered every 7th day. Seedlings appear abundantly after rain, but suffer severely from damage from fungi, insects, rodents, browsing ungulates, fire, *etc.* Some grass or mulch cover seems optimal for establishment of seedlings and cultivation seems beneficial. Seedling survival best when browsed the first two weeks due to compensatory effects of the cotyledons or after 8 weeks when roots are established. Nitrogen fixing.

References. — Kaul and Manohar 1966; Lamprey 1967; Obeid and Seif el Din 1970; Seif el Din and Obeid 1971a, b; Cheema and Quadir 1973; Gupta *et al.* 1973; Tewari and Rathore 1973; Dyson and Thogo 1976; Nongonierma 1978; Muthana and Arora 1980; Allen and Allen 1981; Doran *et al.* 1983; Tonder 1985; Maydell 1986; Tybirk unpublished.

Acacia seyal Del. (MIMOSACEAE)

Dispersal. — The dehiscent pod with the seeds hanging in the funicles is dispersed by wind, birds, and probably by ungulates.

Seed characteristics. — Seeds flattened, often exposed when still green, average 6.6/pod, 39–50 g/1000 seeds. Parenchyma layer 225–250 μm thick.

Seed pests. — 12–16% of seeds in stores have been reported infested. Percentages in parenthesis are examples of known infestation rates for the individual species: *Bruchus elnairensis, Bruchidius aurivillii, B. sahelicus* Decelle (17–96%), *B. sieberianae* Decelle (10–96%), *B. voltaicus* Decelle, *Bruchidius sp., Caryedon capicola* (Motschoulsky) (10%), *C. excavatus* Decelle (96%), *C. mauritanicus* Decelle (96%), *C. sahelicus* Decelle (16–96%).

Germination. — *Acacia seyal* has been considered a medium dormancy species with 26–40% of untreated seeds germinating in 8 weeks, while 32–88% of scarified seeds germinated in 1 week. Boiling and acid treatments have also been applied. Flooding has been reported to initiate germination and trials have revealed that watering every 3rd day is necessary for germination.

Vegetative propagation and seedling growth. — Cuttings and root stocks may be used for regeneration when precipitation is > 500 mm. It coppices when cut before rains and clipping may improve production of shoots and roots, but cutting may kill trees. The growth of seedling is very sensitive to water supply and removal of cotyledons reduces growth and increases mortality. Mean annual height increment of 21 cm has been given but other examples mentions around 75 cm. Nitrogen fixing.

References. — Buchwald 1895; Harker 1959; Nongonierma 1978; Gosseye 1980; Muthana and Arora 1980; Piot 1980; Allen and Allen

1981; Varaigne-Labeyrie and Labeyrie 1981; Bebawi and Mohammed 1982; French Sudan Forest Service cited in Fagg and Greaves 1990a; Rahman and Dafei cited in Fagg and Greaves 1990b; Maydell 1986; Coe and Coe 1987; Schmidt 1988.

Acacia sieberiana DC. (MIMOSACEAE)

Dispersal. — The pod is tardily dehiscent or indehiscent and dispersed primarily by ungulates, and possibly by water.

Seed characteristics. — Large rounded seeds, average 11/pod, 164–222 g /1000 seeds.

Seed pests. — 34% of all stored seeds have been reported infested. Percentages in parenthesis are examples of known infestation rates for the individual species: *Bruchidius baudoni* (Caill.), *B. natalensis* (Pic), *B. centromaculatus*, *B. luteopygus* (Pic) (2–13%), *B. platypennis* Decelle (96%), *B. sahelicus* Decelle (18%), *B. senegalensis* (Pic) (2–81%), *B. sieberianae* Decelle (1–13%), *B. uberatus* (Fahr.) (5%) *Bruchidius* spp., *Caryedon albonotatum* (Pic), *C. interstinctus* (Fahr.) (15%), *Tuberculobruchus natalensis* (Pic) (1–10%).

Germination. — 21.5% of untreated seeds germinated in 20 weeks while 82% of scarified seeds germinated in 2 weeks. Fire induces germination in the upper 2 cm soil and possibly kills developing bruchid larvae in the seeds. Also acid treatments and boiling (1 hour) have been successful.

Vegetative propagation and seedling growth. — No data, but nitrogen fixing.

References. — Buchwald 1895; Prevett 1967; Brown and Booysen 1969; Nongonierma 1978; Allen and Allen 1981; Varaigne-Labeyrie and Labeyrie 1981; Larson cited in Doran *et al.* 1983; Maydell 1986; Sabiiti and Wein 1987.

Acacia tortilis (Forsk.) Hayne (MIMOSACEAE)

Dispersal. — The indehiscent pod is primarily dispersed by ungulates but wind has been mentioned as a possibility.

Seed characteristics. — Rounded, rather small seeds, average 8.2–10.2/pod, 50–66 g/1000 seeds. Parenchyma layer thick, 475–500 μm. Seed coat hardens during maturation.

Seed pests. — 70% of all stored seeds have been found infested by a variety of different bruchids. Percentages in parenthesis are examples of known infestationrates for the individual species: *Bruchus elnairensis, B. albonotatus, Bruchidius albonotatus, B. albosparsus* Fahr., *B. aurivillii* Blanc (11–100%), *B. cadei* Decelle (99%), *B. cadenati* (Pic), *B. centromaculatus, B. latevalvus* Decelle, *B. petechialis* (Gyllenhal), *B. rubicundus* (Fahr), *B. sahelicus* Decelle(13–100%), *B. silaceus* Fahr. (36%), *B. sinaitus* Daniel (36–100%), *B. spadicus* (Fahr.), *B. uberatus* (Fahr.) (22%), *Caryedon capicola* (Motschoulsky) (34%), *C. gonagra, C. longispinosa* Decelle (31–100%), *C. sahelicus* Decelle (7–100%), *Spermophagus rufonotatus, Oedaule sp., Metriocharis silvestris.*

Germination. — 35% of untreated seeds germinated in 15 weeks, while 81.5% of scarified seeds germinated in 2 weeks. Acid treatment (20–120 minutes), dry heat and boiling have also been successful. Flooding can initiate germination and it is dependent on water retention capacity of the growth medium. Germination without pretreatment occurs in spurts during 450 days. Often seen germinate from animal feces. Sowing must be in the upper 4 cm soil.

Vegetative propagation and seedling growth. — Cutting in early rainy season induces coppicing with twice as fast growth as seedling growth. Growth of seedlings depends on soil water potential and nutrient content. No salt tolerance. Removing cotyledons reduces growth. Seedlings depend on full sunlight; shade limits shoot and root biomass, root length and leaf ares. Significant impact of grass competition on seedling growth. Older seeds have been mentioned to give more healthy seedlings than fresh seeds. Root growth >1 m in 3 months and shoot growth in manure/sand mixture can be from 65–130 cm /year. Nitrogen fixing.

References. — Buchwald 1895; Lamprey 1967; Gupta *et al.* 1973; Halevy 1974; Lamprey *et al.* 1974; Karschon 1975; Muthana and Arora 1980; Nongonierma 1978; Pathak *et al.* 1980; Allen and Allen 1981; Doran *et al.* 1983; Karschon cited in Fagg and Greaves 1990b; Bosshard cited in Fagg and Greaves 1990b; Depierre and Gillet cited in Fagg and Greaves 1990b; Rahman and Dafei cited in Fagg and Greaves 1990b; Pellew and Southgate 1984; Bebawi and Mohammed 1985; Gupta and Muthana 1985; Tonder 1985; Coughenour and Detling 1986; Maydell 1986; Ahmed 1987; Coe and Coe 1987; Milton 1988; Schmidt 1988; Knoop cited in Smith and Shackleton 1988; Tolsma 1989; Tybirk unpublished.

Albizia chevalieri Harms (MIMOSACEAE)
Dispersal. — The dehiscent pod is dispersed by wind and possibly by ungulates.
Seed characteristics. — Flattened seeds, 77 g/1000 seeds.
Seed pests. — No data.
Germination. — No data.
Vegetative propagation and seedling growth. — No data. Probably nitrogen fixing.
References. — Buchwald 1895; Allen and Allen 1981; Maydell 1986.

Albizia lebbeck (L.) Benth. (MIMOSACEAE)
Dispersal. — The dehiscent pods rattle in the wind in the dry season. Dispersed by wind and possibly by ungulates.
Seed characteristics. — 100–130 g/1000 seeds.
Seed pests. — *Bruchus submaculatus* (Fahr), *B. tougourensis* Decelle. Predation less than 10%, *B. tougourensis* alone 4%.
Germination. — 60% of untreated seeds germinated in 8 weeks, while 80% of scarified seeds germinated in 2 weeks.

Vegetative propagation and seedling growth. — Cuttings have been successful and it coppices after cutting. Seedlings tolerate weak salt and sodium concentrations but are rather slow growing (14 cm/year). Nitrogen fixing.
References. — Buchwald 1895; Muthana and Arora 1980; Allen and Allen 1981; Varaigne-Labeyrie and Labeyrie 1981; Tomar and Yadav cited in Fagg and Greaves 1990b; Tomar and Gupta 1985; Willan 1985; Babeley *et al.* 1986; Maydell 1986; Ngulube and Chipompha 1987.

Bauhinia rufescens Lam. (CAESALPINIACEAE)
Dispersal. — The indehiscent pod persists on the tree and is eaten and dispersed by ungulates.
Seed characteristics. — 100–111 g/1000 seeds.
Seed pests. — *Caryedon gonagra* (F.), *C. cassieae* (Gyll.), *C. serratus* (Ol.).
Germination. — Boiling has been recommended for rapid germination.
Vegetative propagation and seedling growth. — Rootstocks possible when precipitation is > 500 mm/year. Probably not nitrogen fixing.
References. — Prevett 1965, 1967; Gosseye 1980; Allen and Allen 1981; Maydell 1986.

Cassia siamea Lam. (CAESALPINIACEAE)
Dispersal. — The dehiscent pod is dispersed by wind and probably also by ungulates.
Seed characteristics. — Flattened small seeds, 25–28 g/1000 seeds.
Seed pests. — No data.
Germination. — Young seeds need no pretreatment while old seeds may be either boiled, acid treated or scarified to give rapid germination.
Vegetative propagation and seedling growth. — Has been reported to regenerate from stumps by coppicing. Seedlings develop both tap root and lateral roots from the start. They are difficult to transplant and suffer from grass competition, but can survive moderate salinity. Quite fast growth (up to 8.5m in 4 years). Not known to be nitrogen fixing.
References. — Anonymous 1960; Allen and Allen 1981; Tomar and Gupta 1985; Willan 1985; Maydell 1986; Kariuki 1987.

Cassia sieberiana DC. (CAESALPINIACEAE)
Dispersal. — The indehiscent pod is dispersed by ungulates.
Seed characteristics. — Round seeds, 61–141 g/1000 seeds. Parenchyma layer 285–300 μm thick and very thick cuticula (18–20 μm).
Seed pests. — Rather low predations rates by *Caryedon cassieae* (Gyll.), *C. gonagra* (F.), *C. serratus* (Ol.).
Germination. — Both acid (10–30 minutes gives 90–95% germination in 1 week) treatment and boiling has been recommended, and direct seeding is possible.

Vegetative propagation and seedling growth. — Coppicing abilities have been reported. Nitrogen fixing is found in half of the genus.
References. — Anonymous 1960; Prevett 1965, 1967; Allen and Allen 1981; Varaigne-Labeyrie and Labeyrie 1981; Maydell 1986; Schmidt 1988.

Dalbergia melanoxylon Guill. et Perrott. (FABACEAE)
Dispersal. — The papery pod is dispersed by wind and possibly also by ungulates.
Seed characteristics. — 1–2 flattened seeds/ pod, 63 g/1000 seeds.
Seed pests, germination, vegetative propagation and seedling growth. — No data. Nitrogen fixing.
References. — Buchwald 1895; Allen and Allen 1981; Maydell 1986.

Dichrostachys cinerea (L.) Wight et Arn. (MIMOSACEAE)
Dispersal. — The pods are eaten by ungulates, but also humans and wind have been mentioned as possible dispersal agents.
Seed characteristics. — Large rounded seeds, 313–417 g/1000 seeds.
Seed pests. — Severe predation by the following species: *Bruchidius petechialis* (Gyllenhal), *B. centromaculatus*, *B. dichrostachydis* Decelle, *B. securiger* Decelle, *Bruchidius spp.*, *Spermophagus densepubens*, *S. rufonotatus* , *Metriocharis* nr. *silvestris*.
Germination. — Scarification is recommended for rapid germination.
Vegetative propagation and seedling growth. — Root cuttings and root suckering is possible. Natural seedlings only found in open land, but is rather slow growing (22 cm/year). Nitrogen fixing.
References. — Buchwald 1895; Prevett 1967; Muthana and Arora 1980; Allen and Allen 1981; Willan 1985; Högberg 1986; Maydell 1986; Smith and Goodman 1986; Tolsma 1989.

Entada africana Guill. et Perrott. (MIMOSACEAE)
Dispersal. — The indehiscent segmenting pod is dispersed by wind.
Seed characteristics. — 250 g/1000 seeds.
Seed pests. — No data on infestation.
Germination. — No data.
Vegetative propagation and seedling growth. — Good coppicing abilities after fire. Probably nitrogen fixing.
References. — Brookman-Amissah *et al.* 1980; Allen and Allen 1981; Varaigne-Labeyrie and Labeyrie 1981; Maydell 1986.

Erythrina senegalensis DC. (FABACEAE)
Dispersal. — The pod is moniliform with bright red seeds probably dispersed by birds.
Seed pests. — Seeds described as toxic, but attacks have been recorded.
Germination. — No data.

Vegetative propagation and seedling growth. — Easily propagated by cuttings. Probably nitrogen fixing.
References. — Janzen 1971; Allen and Allen 1981; Maydell 1986.

Leucaena leucocephala (Lam.) de Wit (MIMOSACEAE)
Dispersal. — The pods are dehiscent and probably dispersed both by wind and ungulates.
Seed characteristic. — 45 g/1000 seeds.
Seed pests. — No data on infestation.
Germination. — Boiling has been recommended but direct seeding seems sufficient.
Vegetative propagation and seedling growth. — Good coppicing abilities. Seedling growth best from larger seeds but is limited by water stress. VA-mycorrhizal fungi are very important for water and nutrient uptake. Survives both high salinity and high moisture content. Nitrogen fixing.
References. — Allen and Allen 1981; Pathak *et al.* 1981; Varaigne-Labeyrie and Labeyrie 1981; Tomar and Gupta 1985; Maydell 1986; Osman cited in Milton 1988; Michelsen 1989; Michelsen and Rosendahl 1990.

Mimosa pigra L. (MIMOSACEAE)
Dispersal. — The densely bristled segmenting indehiscent pods are dispersed both by wind, water and by adhering to animals.
Seed characteristics. — 23 g/1000 seeds.
Seed pests. — No infestation reported, but needed as the species has become a weed in some areas.
Germination. — No pretreatment needed.
Vegetative propagation and seedling growth. — No data. Nitrogen fixing.
References. — Buchwald 1895; Ridley 1930; Allen and Allen 1981; Varaigne-Labeyrie and Labeyrie 1981; Pijl 1982; Maydell 1986.

Parkia biglobosa (Jacq.) Benth. (MIMOSACEAE)
Dispersal. — The woody indehiscent pod or seeds are eaten by monkeys, humans and ungulates which all may serve as dispersal agents.
Seed characteristics. — 200 g/1000 seeds.
Seed pests. — No data on infestation.
Germination. — Boiling has been recommended.
Vegetative propagation and seedling growth. — Root suckers have been reported. Seedlings minimize drought impact by increasing root/shoot ratio, and only open stomata in the morning using stored water when stomata close. Probably nitrogen fixing.
References. — Buchwald 1895; Allen and Allen 1981; Varaigne-Labeyrie and Labeyrie 1981; Hopkins 1983; Hopkins and White 1984; Maydell 1986; Osunubi and Fasehun 1987.

Parkinsonia aculeata L. (CAESALPINIACEAE)
Dispersal. — The pods are consumed by ungulates, and humans have also been reported as dispersers.
Seed characteristics. — Round elongate seeds, 83 g/1000 seeds.
Seed pests. — *Caryedes germanii* and *Mimosestes amicus* (Horn) are known to infest seeds in the native range but no infestation has been reported from Sahel.
Germination. — 50% of untreated seeds germinated in 3 weeks, while 100% germinated in 7 days after scarification. Also boiling water and 20 minutes acid treatment are recommended.
Vegetative propagation and seedling growth. — Root and shoot cuttings have been reported to be successful. Seedlings tolerate weak salt and sodium concentrations. Not nitrogen fixing.
References. — Buchwald 1895; Center and Johnson 1974; Allen and Allen 1981; Varaigne-Labeyrie and Labeyrie 1981; Tomar and Yadav cited in Fagg and Greaves 1990b; Tomar and Gupta 1985; Willan 1985; Maydell 1986; Ngulube and Chipompha 1987.

Piliostigma reticulatum (DC.) Hoechst. (CAESALPINIACEAE)
Dispersal. — The woody indehiscent pod is dispersed by ungulates and perhaps water.
Seed characteristics. — Rounded seeds, 69–91 g/1000seeds.
Seed pests. — *Caryedon serratus* (Ol.) has been reported to attack 9% of the seeds.
Germination. — Hot water has been recommended for hastening germination.
Vegetative propagation and seedling growth. — No data. Probably not nitrogen fixing.
References. — Prevett 1965, 1966; Allen and Allen 1981; Varaigne-Labeyrie and Labeyrie 1981; Maydell 1986.

Piliostigma thonningii (Schum.) Milne-Redh. (CAESALPINIACEAE)
Dispersal. —The woody indehiscent pod is eaten by ungulates.
Seed characteristics. — Rounded seeds, 118 g/1000 seeds.
Seed pests. — *Caryedon serratus* attacks the seeds.
Germination. — Soaking for 24 hours in water hastens germination.
Vegetative propagation and seedling growth. — Root suckers are known to occur. Not nitrogen fixing.
References. — Prevett 1965, 1966, 1967; Allen and Allen 1981; Varaigne-Labeyrie and Labeyrie 1981; Maydell 1986.

Prosopis africana (Guill., Perrott. et Rich.) Taub. (MIMOSACEAE)
Dispersal. — The woody indehiscent pod is probably eaten by ungulates and primates.
Seed characteristics. — 125–153 g/1000 seeds.
Seed pests. — *Caryedon cassieae* (Gyll.).

Germination. — For rapid germination hot water has been recommended.
Vegetative propagation and seedling growth. — No data. Probably nitrogen fixing.
References. — Buchwald 1895; Prevett 1965; Anonymous 1979; Allen and Allen 1981; Maydell 1986.

Prosopis juliflora (SW.) DC. (MIMOSACEAE)
Dispersal. — The indehiscent pod is eaten by ungulates, but also humans have been mentioned as dispersal agents.
Seed characteristics. — 67–114 g/1000 seeds.
Seed pests. — Known predators in native range: *Acanthoscelides* sp., *Amblycerus epsilon, A. prosopis, Algarobius bottimori, Caryedon* sp., *Mimosestes amicus* (Horn), *M. insularis* Kingsolver and Johnson, *M. protractus* Horn, *Rhipobruchus prosopis, R. psephenopygus, Scutobruchus ceratioborus* (Philippi).
Germination. — 40–60% of untreated seeds will germinate within few weeks, while 90% of scarified seeds will germinate in one week. Also shaking in container, boiling and acid treatments have been recommended. Direct sowing before rains can be worth while. Germination is hampered by grass cover.
Vegetative propagation and seedling growth. — Shoot or branch cuttings successful and root suckering has been reported. Seedlings tolerate moderate to high salinity and sodium concentrations. Browsing by ungulates only on the youngest shoots but attacked severely by termites and rats. May grow up to 2–3 m in one year without watering or protection, but others report 46–66 cm as mean annual height increment. Nitrogen fixing.
References. — Kaul 1956; de Candolle 1962; Mooney *et al.* 1977; Muthana and Arora 1980; Pedersen 1980; Allen and Allen 1981; Tomar and Yadav cited in Fagg and Greaves 1990b; Tomar and Gupta 1985; Maydell 1986.

Pterocarpus erinaceous Poir. (FABACEAE)
Dispersal. — The winged indehiscent pod with long bristles is dispersed by wind but is perhaps also dispersed epizooically.
Seed characteristics. — 286 g/1000 seeds.
Seed pests. — No data on infestation.
Germination, vegetative propagation and seedling growth. — No data, but probably nitrogen fixing.
References. — Buchwald 1895; Allen and Allen 1981; Varaigne-Labeyrie and Labeyrie 1981; Maydell 1986.

Pterocarpus lucens Lepr. ex Guill. et Perrott. (FABACEAE)
Dispersal. — The papery indehiscent pod is typically dispersed by wind but probably also by ungulates.
Seed characteristics. — 200 g/1000 seeds.

Seed pests and germination. — Boiling gave no germination.
Vegetative propagation and seedling growth. — Coppices. Root stocking possible when precipitation is > 500 mm. Probably nitrogen fixing.
References. — Buchwald 1895; Gosseye 1980; Allen and Allen 1981; Willan 1985; Maydell 1986.

Tamarindus indica L. (CAESALPINIACEAE)

Dispersal. — The indehiscent pod including the pulp is eaten by ungulates, monkeys, humans, and birds.
Seed characteristics. — Large rounded seeds, 400–500 g/1000 seeds.
Seed pests. — *Caryedon serratus* may be a serious pest.
Germination. — Fresh seeds germinate directly (92% in 5 weeks) while older seeds need boiling or scarification (100% germination in 3 weeks).
Vegetative propagation and seedling growth. — Branch cuttings have been described as a possibility. Not nitrogen fixing.
References. — Buchwald 1895; Prevett 1965, 1966, 1967; Allen and Allen 1981; Varaigne-Labeyrie and Labeyrie 1981; Maydell 1986; Ngulube and Chipompha 1987.

8. LITERATURE CITED

Ahmed El Houri, A. 1986. Some aspects of dry land afforestation in the Sudan with special reference to *Acacia tortilis* (Forsk.) Hayne, *A. senegal* Willd. and *Prosopis chilensis* (Molina) Stuntz. — Forest Ecology and Management 16: 209–221.

Ahmed, P. 1987. Vegetative propagation of *Acacia albida*: A promising species for farm-forestry in arid areas. — Indian Forester 113: 459–465.

Allen, O. N. and Allen, E. K. 1981. The Leguminosae. A source book of characteristics, uses and nodulation. — Macmillan, London.

Anonymous 1960. *Cassia* spp. Caractères sylvicoles et méthodes de plantation. — Bois et Fôrets des Tropiques 70: 43–48.

Anonymous 1979. Fifty trees of Abuko Nature reserve. — Conservation Fact Sheet 9: 1–12.

Anonymous 1988. *Faidherbia albida* (Del.) A. Chev. (Synonyme: *Acacia albida* Del.). Monographie. — Centre Technique Forestier Tropical, Nogent-sur-Marne.

Augspurger, C. K. 1989. Morphology and aerodynamics of wind dispersed legumes. — In: Stirton, C. H. and Zarucci, J. L. (Eds.), Advances in legume biology. Monogr. Syst. Bot. Missouri Bot. Gard 29: 451–466.

Babeley, G. S., Gautam, S. P. and Kandya, A. K. 1986. Pretreatment of *Albizia lebbeck* Benth. seeds to obtain better germination and vigour. — J. Tropical Forestry 2(2): 105–113.

Bakolimalala, R. 1989. Malagasy Leguminosae: Potentials for fuelwood and reforestation. Preliminary results. — Pp. 167 – 182 in: International Foundation for Science (Eds.), Proceedings of a regional seminar on trees for development in Sub-Saharan Africa, February 20–25 1989. IFS, Stockholm.

Bebawi, F. F. and Mohammed, S. M. 1982. Effects of irrigation frequency on germination and on root and shoot yields of *Acacia* species. — Plant and Soil 65(2): 275–279.

Bebawi, F. F. and Mohamed, S. M. 1985. The pretreatment of seeds of six Sudanese acacias to improve their germination response. — Seed Sci. and Technol. 13: 111–119.

Belinsky, A. and Kugler, J. 1978. Observations on the biology and host preference of *Caryedon serratus palaestinicus* (Coleoptera: Bruchidae) in Israel. — Israel J. Entomology 12: 19–33.

Bhatnagar, S. P. and Johri, B. M. 1972. Development of angiosperm seeds. — Pp. 78–149 in: Kozlowski, T. T. (Ed.), Seed biology. Academic Press, New York.

Brenan, J. P. M. 1959. Leguminosae, Mimosoideae. — In: Hubbard, C. E. and Milne-Redhead, E. (Eds.), Flora of Tropical East Africa. Crown Agent for Overseas Government and Administration.

Brookman-Amissah, J., Hall, J. B., Swaine, M. D. and Attakorah, J. Y. 1980. A reassessment of a fire protection experiment in north-eastern Ghana savanna. — J. Appl. Ecol. 17: 85–99.

Brown, N. A. C. and Booysen, P. de van 1969. Seed coat impermeability of several *Acacia* species. — Agroplantae 1: 51–60.

Buchwald, J. 1895. Die Verbreitungsmittel der Leguminosen des Tropisches Afrika. — Bot. Jahrbücher 19: 494–561.

Burtt, B. D. 1929. A record of fruits and seeds dispersed by mammals and birds from the Singida district of Tanganyika territory. — J. Ecol. 17(2): 351–355.

Cavanagh, A. K. 1980. A review of some aspects of germination of acacias. — Proc. Roy. Soc. Victoria. 91: 161–180.

Center, T. D. and Johnson, C. D. 1974. Coevolution of some seed beetles (Coleoptera: Bruchidae) and their hosts. — Ecology 55: 1096–1103.

Cheema, M. S. Z. A. and Quadir, S. A. 1973. Autecology of *Acacia senegal* (L.) Willd. — Vegetatio 27(1–3): 131–162.

Coe, M. and Coe C. 1987. Large herbivores, acacia trees and bruchid beetles. — S. Afr. J. Sci. 83: 624–635.

Coughenour, M. B. and Detling, J. K. 1986. *Acacia tortilis* seed germination responses to water potential and nutrients. — Afr. J. Ecol. 24: 203–205.

De Candolle, A. 1962. *Prosopis juliflora*. Charactères sylvicole et Méthodes de plantation. — Bois et Fôrets des Tropiques 82: 33–38.

Decelle, J. E. 1951. Contribution á l'étude des Bruchidae du Congo Belge (Col. Phytophaga). — Rev. Zool. Bot. Afr. 45(1–2): 172–192.

Dell, B. 1980. Structure and function of the strophiolar plug in seeds of *Albizia lophantha*. — Amer. J. Bot. 67(4): 556–563.

Donahaye, E., Navarro, S. and Calderon, M. 1966. Observation on the life cycle of *Caryedon gonagra* (F) on its natural host in Israel, *Acacia spirocarpa* and *A. tortilis*. — Tropical Science 8: 85–89.

Doran, J. C., Turnbull, J. W. Boland, D. J. and Gunn, B. V. 1983. Handbook on seeds of dry-zone acacias. — FAO, Rome.

Duke, J. A. and Polhill, R. M. 1981. Seedlings of the Leguminosae. — Pp. 941–950 in: Polhill, R. M. and Raven, P. H. (Eds.), Advances in legume systematics. Royal Botanical Garden, Kew.

Dyson, W. G. and Thogo, S. 1976. The growth of *Acacia* Mill. at Muguga arboretum 1952–1975. Forestry Technical Note. — East African Agriculture and Forestry Research Organisation, Muguga.

Ellner, S. and Shmida, A. 1981. Why are adaptations for long-range seed dispersal rare in desert plants? — Oecologia 51: 133–144.

Ernst, W. H. O., Tolsma, D. J. and Decelle, J. E. 1989. Predation of seeds of *Acacia tortilis* by insects. — Oikos 54: 294–300.

Fagg, C. W. and Greaves, A. 1990a. *Acacia nilotica.* — Annotated bibliography, F 42, Commenwealth Agricultural Bureau, Oxon, U.K.

Fagg, C. W. and Greaves, A. 1990b. *Acacia tortilis.* — Annotated bibliography, F 41, Commenwealth Agricultural Bureau, Oxon, U.K.

Farrel, T. P. and Ashton, D. H. 1978. Population studies on *Acacia melanoxylon* R. Br. 1. Variation in seed and vegetative characteristics. — Aust. J. Bot. 26: 365–379.

Felker, P. 1978. State of the art: *Acacia albida* as a complementary permanent intercrop with annual crops. USAID report. — University of California, Riverside.

Fenner, M. 1985. Seed ecology. — Chapman and Hall, London.

Gassama, Y. K. 1989. Culture in vitro et amélioration symbiotique chez *Acacia albida* (Leguminosae) adulte. — Pp. 286–290 in: International Foundation for Science (Eds.), Proceedings of a regional seminar on trees for development in Sub-Saharan Africa, February 20–25 1989. IFS, Stockholm.

Giffard, P. L. 1964. Les possibilités de reboisement en *Acacia albida* au Sénégal. — Bois et Forêts des Tropiques 95: 21–33.

Giffard, P. L. 1966. Les Gommiers: *Acacia senegal* Willd., *Acacia laeta* R. Br. — Bois et Forêts des Tropiques 105: 21–32.

Giffard, P. L. 1971. Recherches complémentaires sur *Acacia albida* (Del.). — Bois et Forêts des Tropiques 135: 3–20.

Gosseye, P. 1980. Introduction of browse plants in the Sahelo-Sudanian zone. — Pp. 393–397 in: Le Houérou, H. N. (Ed.), Browse in Africa. ILCA, Addis Abeba.

Grice, A. C. and Westoby, M. 1987. Aspects of the dynamics of the seed-banks and seedling populations of *Acacia victoriae* and *Cassia* spp. in arid western New South Wales. — Australian J. Ecol. 12: 209–215.

Gunn, C. R. 1981. Fruits and seeds in the subfamily Mimosoideae (Fabaceae). Technical Bulletin. no. 1681 — U.S. Department of Agriculture, Springfield.

Gupta, J. P. and Muthana, K. D. 1985. Effect of integrated moisture conservation technology on the early growth and establishment of *Acacia tortilis* in the Indian desert. — Indian Forester 111: 477–485.

Gupta, R. K., Saxena, S. K. and Dutta, B. K. 1973. Germination, seedling behavior and phytomass of some acacias in the nursery stage. — Indian Forester 99: 352–358.

Gwynne, M. D. 1969. The nutritive value of *Acacia* pods in relation to *Acacia* seed distribution by ungulates. — E. Afr. Wildlife J. 7: 176–178.

Habish, H. A. 1970. Effect of certain soil conditions on nodulation of *Acacia* spp. — Plant and Soil 33: 1–6.

Halevy, G. 1974. Effects of gazelles and seed beetles (Bruchidae) on germination and establishment of *Acacia* species. — Israel J. Bot. 23: 120–126.

Halevy, G. and Orshan, G. 1972. Ecological studies on *Acacia* species in the Negev and Sinai. I. Distribution of *Acacia raddiana, A. tortilis* and *A. gerrardii* ssp. *negevensis* as related to environmental factors. — Israel J. Bot. 21: 197–208.

Hanna, P. J. 1984. Anatomical features of the seed coat of *Acacia kempeana* (Mueller) which relate to increased germination rate induced by heat treatment. — New Phytol. 96: 23–29.

Hanna, P. J. and Burridge, T. 1983. Features of *Acacia* seed coats which relate to germination. — Bull. Int. Group Study Mimosoideae 11: 63–67.

Harker, K. W. 1959. An *Acacia* weed of Uganda grasslands. — Tropical Agriculture 36: 45–51.

Harper, J. L. 1977. Population biology of plants. — Academic Press, London.

Harvey, G. J. 1981. Recovery and viability of Prickly Acacia (*Acacia nilotica* ssp. *indica*) seed ingested by sheep and cattle. — Pp. 197–201 in: Wilson, B. J. and Swarbrick, J. T. (Eds.), Proceedings of the sixth Australian weeds conference. Queensland Weed Society.

Hayward, D. F. and Oguntoyinbo, J. S. 1987. The climatology of West Africa. — Hutchinson, London.

Hoffmann, M. T., Cowling, R. M., Douie, C. and Pierce, S. M. 1989. Seed predation and germination of *Acacia erioloba* in the Kuiseb River Valley, Namib Desert. — S. Afr. J. Bot. 55(1): 103–106.

Hopkins, H. 1983. The taxonomy, reproductive biology and economic potential of *Parkia* (Leguminosae: Mimosoideae) in Africa and Madagascar. — Bot. J. Linn. Soc. 87: 135–167.

Hopkins, H. C. and White, F. 1984. The ecology and chorology of *Parkia* in Africa. — Bull. Jard. Bot. Nat. Belg. 54: 235–266.

Howe, R. W. 1972. Insects attacking seeds during storage. — Pp. 247–300 in: Kozlowski, T. T. (Ed.), Seed biology. Academic Press, New York.

Howe, H. F. and Smallwood, J. 1982. Ecology of seed dispersal. — Ann. Rev. Ecol. Syst. 13: 201–228.

Hutchinson, J. and Dalziel, J. M. 1954–1958. Flora of tropical West Africa. — Whitefriars Press Ltd., London.

Hyde, E. O. C. 1954. The function of the hilum in some Papilionaceae in relation to the ripening of the seed and the permeability of the testa. — Ann. Bot. 18: 241–256.

Högberg, P. 1986. Soil nutrient availability, root symbioses and tree species composition in tropical Africa: a review. — J. Trop. Ecol. 2: 359–372.

Högberg, P. 1989. Root symbiosis of trees in savannas. — Pp. 121–136 in: Proctor, J. (Ed.), Mineral nutrients in tropical forest and savanna ecosystems. Blackwell Scientific Publications, Oxford.

Janzen, D. H. 1969. Seed-eaters versus seed size, number, toxicity and dispersal. — Evolution 23(1): 1–27.

Janzen, D. H. 1971. Seed predation by animals. — Ann. Rev. Ecol. Syst. 2: 465–492.

Janzen, D. H. 1986. Mice, big mammals, and seeds: it matters who defecates what where. — Pp. 251–271 in: Estrada, A. and Fleming, T. H. (Eds.), Frugivores and seed dispersal. Junk Publishers, Dordrecht.

Jarman, P. J. 1976. Damage to *Acacia tortilis* seeds eaten by impala. — E. Afr. Wildl. J. 14: 223–225.

Johnson, C. D. 1981. Seed beetle host specificity and the systematics of the Leguminosae. — Pp. 995–1027 in: Polhill, R. M. and Raven, P. (Eds.), Advances in legume systematics. Royal Bot. Gard., Kew.

Johnson, C. D. 1983. Handbook on seed insects of *Prosopis* species. — FAO, Rome.

Kariuki, E. M. 1987. Effects of presowing treatments on seed germination of four important tree species in Kenya. — Pp. 143–149 in: Kamra, S. K. and Ayling, R. D. (Eds.), Proceedings of the international symposium on forest seed problems in Africa. IUFRO and Swedish Univ. Agric. Sci., Umeå.

Karschon, R. 1975. Seed germination of *Acacia raddiana* Savi and *A. tortilis* Hayne as related to infestation by Bruchids. — Agricultural Research Organisation Leaflet 52: 1–8.

Karschon, R. 1976. Clonal growth pattern of *Acacia albida* Del. — Bull. Int. Group Study Mimosoideae 4: 28–30.

Kaul, O. N. 1956. Propagating mesquite by root and shoot cuttings. — Indian Forester 82(11): 569–572.

Kaul, R. N. and Manohar, M. S. 1966. Germination studies on arid zone tree seeds. — Indian Forester 92(8): 499–503.

Khan, M. A. W. 1970. Phenology of *Acacia nilotica* and *Eucalyptus microtheca* at wadi Medani (Sudan). — Indian Forester 96: 226–248.

Kingsolver, J. M., Johnson, C. D., Swier, S. R. and Teran, A. 1977. *Prosopis* fruits as a resource for invertebrates. — Pp. 108–122 in: Simpson, B. B. (Ed.), Mesquite — Its biology in two desert ecosystems. Halsted Press, Stroudsbourg.

Lamprey, H. F. 1967. Notes on the dispersal and germination of some tree seeds through the agency of mammals and birds. — E. Afr. Wildl. J. 5: 179–180.

Lamprey, H. F., Halevy, G. and Machaka, S. 1974. Interactions between *Acacia*, bruchid seed beetles and large herbivores. — E. Afr. Wildl. J. 12: 81–85.

Leistner, O. A. 1961. On the dispersal of *Acacia giraffae* by game. — Koedoe 4: 101–104.

Luca, Y. de 1965. Cataloque des metazoaires parasites et predateurs de bruchides (Coleoptera). — J. Stored Products Res. 1: 51–98.

Mathur, R. S., Sharma, K. K. and Rawat, M. M. S. 1984. Germination behaviour of various provenances of *Acacia nilotica* ssp. *indica*. — Indian Forester 110: 435–449.

Maydell, H.-J. von 1986. Trees and shrubs of the Sahel. Their characteristics and uses. — GTZ Verlagsgesellschafft, Rossdorf.

Michelsen, A. 1989. Effects of vesicular arbuscular mychorrizal fungi and drought stress on growth on *Acacia nilotica* and *Leucaena leucocephala* seedlings. — Unpublished M.Sc. Thesis, Inst. Botanical Ecology, Copenhagen University.

Michelsen, A. and Rosendahl, S. 1990. The effect of VA mycorrhizal fungi, phospohrous and drought stress on the growth of *Acacia nilotica* and *Leucaena leucocephala* seedlings. — Plant and Soil 124: 7–13.

Milton, S. J. 1988. The effects of pruning on shoot production and basal increment of *Acacia tortilis*. — S. Afr. J. Bot. 54(2): 109–117.

Milton, S. J. and Hall, A. V. 1981. Reproductive biology of Australian acacias in the South-Western Cape province, South Africa. — Trans. Roy. Soc. S. Afr. 44(3): 465–485.

Mooney, H. A., Simpson, B. B. and Solbrig, O. T. 1977. Phenology, morphology, physiology. — Pp. 26–43 in: Simpson, B. B. (Ed.), Mesquite — its biology in two desert ecosystems. Halsted Press, Stroudsbourg.

Muthana, K. D. and Arora, G. D. 1980. Performance of *Acacia tortilis* (Forsk.) under different habitats of the Indian arid zone. — Ann. of Arid Zone 19(1/2): 110–118.

New, T. R. 1983. Seed predation of some Australian acacias by weevils (Coleoptera: Curculionidae). — Aust. J. Zool. 31: 345–352.

Ngulube, M. R. and Chipompha, N. W. S. 1987. Effects of seed pretreatment on the germination of some hardwood species for dry zone afforestation in Malawi. — Pp. 216–224 in: Kamra, S. K. and Ayling, R. D. (Eds.), Proceeding of the international symposium on forest seed problems in Africa. IUFRO and Swedish Univ. Agric. Sci., Umeå.

Nongonierma, A. 1978. Contribution à l'étude biosystématique du genre *Acacia* Miller en Afrique occidentale. Thése de Doctorat de l'Etat, 3 Tomes. Faculté des Sciences de l'Universite de Dakar.

O'Dowd, D. J. and Gill, A. M. 1986. Seed dispersal syndromes in Australian *Acacia*. — Pp. 87–121 in: Murray, D. R. (Ed.), Seed dispersal. Academic Press, Sydney.

Obeid, M. and Seif el Din, A. 1970. Ecological studies of the vegetation of the Sudan. I. *Acacia senegal* (L.) Willd. and its natural regeneration. — J. Appl. Ecol. 7: 507–518.

Obeid, M. and Seif el Din, A. 1971. Ecological studies of the vegetation of the Sudan. III. The effect of simulated rainfall distribution at different isohyets on the regeneration of *Acacia senegal* (L.) Willd. on clay and sandy soils. — J. Appl. Ecol. 8: 203–209.

Olsson, L. 1985. An integrated study of desertification. — Lund Studies in Geography 13: 1–170.

Osunubi, O. and Fasehun, F. E. 1987. Adaptations to soil drying in woody seedlings of African locust bean, (*Parkia biglobosa* (Jacq.) Benth.). — Tree Physiology 3: 321–330.

Pathak, P. S., Gupta, S. K. and Roy, R. D. 1980. Studies on seed polymorphism, germination and seedling growth of *Acacia tortilis* Hayne. — Ind. J. For. 3(1): 64–67.

Pathak, P. S., Rai, M. P. and Debroy, R. 1981. Seed weight affecting early seedling growth attributes in *Leucaena leucocephala* (Lam.) De Wit. — Van Vigyan 19(3): 97–101.

Peake, F. G. G. 1952. On a Bruchid seed-borer in *Acacia arabica*. — Bull. Entomol. Res. 43: 317–324.

Pedersen, B. O. 1980. A note on the genus *Prosopis*. — Int. Tree Crops J. 1: 113–123.

Pellew, R. A. and Southgate, B. J. 1984. The parasitism of *Acacia tortilis* seeds in the Serengeti. — Afr. J. Ecol. 22: 73–75.

Pieterse, P. J. and Cairns, A. L. P. 1986. The effect of fire on an *Acacia longifolia* seed bank in the south-western Cape. — S. Afr. J. Bot. 53: 233–236.

Pijl, L. van der 1982. Principles of dispersal in higher plants. — Springer-Verlag, Berlin.

Piot, J. 1980. Management and utilization methods for ligneous forages: natural stands and artificial plantations. — Pp. 339–349 in: Le Houérou, H. N. (Ed.), Browse in Africa. ILCA, Addis Abeba.

Preece, P. B. 1971. Contributions to the biology of Mulga. II. Germination. — Aust. J. Bot. 19: 39–49.

Prevett, P. F. 1965. The genus *Caryedon* in northern Nigeria with descriptions of six new species (Col.: Bruchidae). — Ann. Soc. Ent. Fr. (N. S.) 1(3): 523–547.

Prevett, P. F. 1966 . Observations on the biology of six species of Bruchidae (Coleoptera) in northern Nigeria. — Entomologists Monthly Magazine 102: 174–180.

Prevett, P. F. 1967. Notes on the biology, food plants and distribution of Nigerian Bruchidae (Coleoptera), with particular reference to the northern region. — Bull. Entomological Soc. Nigeria 1: 3–6.

Quinlivan, B. J. 1966. The relationship between temperature fluctuations and the softening of hard seeds of some legume species. — Aust. J. Agric. Res. 17: 625–631.

Radwanski, S. A. and Wickens, G. E. 1967. The ecology of *Acacia albida* on mantle soils in Zalingei, Jebel Marra, Sudan. — J. Appl. Ecol. 4: 569–579.

Renner, S. S. 1987. Seed dispersal. — Progress in Botany 49: 413–432.

Ridley, H. N. 1930. The dispersal of plants throughout the world. — Reeve and Co. Ltd., Ashford.

Robert, P. 1985. A comparative study of some aspects of the reproduction of three *Caryedon serratus* strains in prescence of its potential host plants. — Oecologia 65: 425–430.

Rolston, M. P. 1978. Water impermeable seed dormancy. — Bot. Review 44: 365–396.

Ross, J. H. 1965. Notes on insect infestation in seed of *Acacia caffra* (Thunb.) Willd. in Natal. — Ann. Natal. Mus. 18: 221–226.

Ross, J. H. 1979. A conspectus of the African *Acacia* species. — Mem. Bot. Surv. S. Afr. 44: 1–155.

Sabiiti E. N. and Wein, R. W. 1987. Fire and *Acacia* seeds: a hypothesis of colonization success. — J. Ecol. 74: 937–946.

Schmidt, L. H. 1988. A study of natural regeneration in transitional lowland rainforest and dry bushland in Kenya. — Unpublished M.Sc. Thesis, Botanical Institute, Aarhus University.

Seif el Din, A. and Obeid, M. 1971a. Ecological studies of the vegetation of the Sudan. II. The germination of seeds and establishment of seedlings of *Acacia senegal* (L.) Willd. under controlled conditions in the Sudan. — J. Appl. Ecol. 8: 191–201.

Seif el Din, A. and Obeid, M. 1971b. Ecological studies of the vegetation of the Sudan. IV. The effect of simulated grazing on the growth of *Acacia senegal* (L.) Willd. seedlings. — J. Appl. Ecol. 8: 211–216.

Silander, J. A. 1978. Density-dependent control of reproductive success in *Cassia biflora*. — Biotropica 10(4): 292–296.

Skaife, S. H. 1926. The bionomics of Bruchidae. — S. Afr. J. Sci. 23: 575–588.

Smith, T. M. and Goodman, P. S. 1986. The effect of competition on the structure and dynamics of *Acacia* savannas in southern Africa. — J. Ecol. 74: 1031–1044.

Smith, T. M. and Shackleton, S. E. 1988. The effects of shading on the establishment and growth of *Acacia tortilis* seedlings. — S. Afr. J. Bot. 54(4): 375–379.

Sniezko, R. A. and Stewart, H. T. L. 1989. Range-wide provenance variation in growth and nutrition of *Acacia albida* seedlings propagated in Zimbabwe. — Forest Ecol. and Management 27: 179–197.

Southgate, B. J. 1975. Seed predation in the genus *Acacia*. — Bull. Int. Group Study Mimosoideae 3: 33–34.

Southgate, B. J. 1978. Variation in the susceptibility of African *Acacia* (Leguminosae) to seed beetle attack. — Kew Bull. 32: 541–544.

Southgate, B. J. 1979. Biology of the Bruchidae. — Ann. Rew. Entomol. 24: 449–473.

Southgate, B. J. 1981. The ecology of bruchids attacking legumes. — Junk Publishers, The Hague.

Southgate, B. J. 1983. Handbook on seed insects of *Acacia* species. — FAO, Rome.

Tewari, M. N. and Rathore, L. S. 1973. Changes in the level of ascorbic acid in germinating seeds of *Acacia senegal* Willd. — Biochem. Physiol. Pflanzen 164: 103–106.

Tolsma, D. J. 1989. On the ecology of savanna ecosystems in south-eastern Botswana. — Centrale Huisdrukkerij Vrije Universiteit, Amsterdam.

Tomar, O. S. and Gupta, R. K. 1985. Performance of some forest tree species in saline soils under shallow and saline water table conditions. — Plant and Soil 87: 329–335.

Tonder, S. J. van 1985. Annotated records of southern African Bruchidae (Coleoptera) associated with acacias, with description of a new species. — Phytophylactica 17: 143–148.

Tran, V. N. 1983. Summary of research on *Acacia* seeds. — Bull. Int. Group Study Mimosoideae 11: 68–75.

Tran, V. N. and Cavanagh, A. K. 1980. Taxonomic implications of fracture load and deformation histograms and the effects of treatments on the impermeable seed coat of *Acacia* species. — Aust. J. Bot. 28: 39–51.

Tybirk, K. 1988. *Acacia nilotica* in Kenya: aspects of flowering, pollination, seed production and regeneration. — Unpublished M.Sc. Thesis, Botanical Institute, Aarhus University.

Tybirk, K. 1989. Flowering, pollination and seed production of *Acacia nilotica*. — Nord. J. Bot. 9: 375–381.

Tybirk, K. 1991. Planting trees in Sahel — doomed to failure? — Pp. 22–28 in: Poulsen, E. and Lawesson, J. E. (Eds.), Dryland degradation: causes and consequenses. Aarhus University Press, Aarhus.

Tybirk, K., Schmidt, L. H. and Hauser, T. (*in prep.*) Soil seed bank of African acacias. — Submitted to Afr. J. Ecol.

Varaigne-Labeyrie, C. and Labeyrie, V. 1981. First data on Bruchidae which attack the pods of legumes in Upper Volta of which eight species are man consumed. — Series Entomologica 19: 83–96.

Venable, D. L. and Brown, J. S. 1988. The selective interactions of dispersal, dormancy, and seed size as adaptations for reducing risk in variable environments. — Amer. Naturalist 131(3): 360–384.

Werker, E. 1980/81. Seed dormancy as explained by the anatomy of embryo envelopes. — Israel J. Botany 29: 22–44.

White, F. 1983. The vegetation of Africa. A descriptive memoir to accompany the UNESCO/AETFAT/UNSO vegetation map of Africa. Unesco, Paris.

Wickens, G. E. 1969. A study of *Acacia albida* Del. (Mimosoïdeae). — Kew Bull. 23: 181–208.

Willan, R. L. 1985. A guide to forest seed handling with special reference to the tropics. — Danida Forest Seed Centre, Humlebæk and FAO, Rome.

Wrangham, R. W. and Waterman, P. G. 1981. Feeding behaviour of Vervet Monkeys on *Acacia tortilis* and *Acacia xanthophloea*: with special reference to reproductive strategies and tannin production. — J. Animal Ecology 50: 715–731.

9. GLOSSARY

aril — an expansion of the funicle enveloping the seed

bush encroachment — the spreading of impenetrable useless thickets of thorny bushes

clonal growth — one individual grows by root suckers to apparently many individuals

conspecific — within species

cotyledons — the first leaves of the embryo; in legume seeds the cotyledons are large, containing nutrition for germination

dehiscent — opening at maturity

diapause — resting phase with almost no biological activity

diaspore — the dispersal unit; either seed, pod, or infructescence

dispersal syndrome — the non random occurrence of combinations of diaspore traits related to the nature of the most probable dispersal agent

embryo — the rudimentary plant formed in a seed

endozoochory — dispersal by passing through the alimentary canal of animals

enforced dormancy — inability to germinate because of environmental constraints

epizooic — dispersal by external adhering to animals

exogenous dormancy — physical impermeability, chemical inhibition or mechanical resistance

granivores — eating seeds

hemi-legume — dehisced papery pod separated in two halfs each with few – many flattened seeds attached

hilum — the site of entrance of funiculus (Figure 4.1)

imbibe — take up water

indehiscent — not opening at maturity

induced dormancy — an acquired condition of inability to germinate caused by some experience after ripening

infructescence — group of fruits from the same inflorescence

innate dormancy — the condition of seeds as they leave the parent plant, viable, but prevented from germinating by certain external factors

lens — an area of weakness in the seed coat where water penetrates (Figure 4.1)

mechanical scarification — to scarify seeds by nicking, filing, drilling, using soldering iron, shaking in container *etc.*

meristem culture — cultivating plant tissue capable of developing into a full plant in laboratory

mesocarp — the often juicy layer between the seeds and the outer layer of a pod

millipedes — terrestrial crustacean with many leg pairs *e.g.* common wood-louse

morphological endogenous — underdevelopment of embryo

moults — periodically shedding of the outer covering, ecdysis

multivoltine — having more lifecycles per year

nitrogenous compounds — here: alkaloids containing nitrogen

nodulation — developing root nodules for nitrogen-fixation

nutritive package — the nutrition found in the cotyledons for germination of the seed

oviposition — laying eggs

palisade layer — columnar outer epidermal cells of the seed coat (Figure 4.2)

phaneroepigeal germination — cotyledons escaping the seed testa aboveground

phenology — the times of recurring natural phenomena especially in relation to climatic conditions

physiological endogenous — with physiological inhibiting mechanism of the full developed embryo

phytophagous — feeding on plants

pleistocene — approximately the last two million years

pleurogram — characteristic line on the seeds of *e.g.* acacias (Figure 4.1)

polymorphism — occurrence of different forms of individuals of the same species

post-dispersal seed predation — seed predation after dispersal

predator satiation — to produce more seeds than the predators can consume

propagules — a shoot capable of developing into an adult

pupal cocoons — the case in which the beetle develop from larva to imago

root suckers — shoots springing from part of roots remote from the main stem

samara — whole or segmented flat papery fruit containing 1 – a few flat seeds

secondary dispersal — dispersal *e.g.* by dung beetle after dispersal by ungulate

seed dormancy — resting phase of seeds

seed predation — the consumption of seeds by animals killing or severely reducing viability of the seed

strophiole — see lens

suberization — to become corky

tumbleweeds — plants being dispersed by tumbling of the fruit

ungulate — hoofed mammal

univoltine — having one cycles per year

VA-mycorrhizal — Vescicular-Arbuscular-Mychorrhiza, a symbiotic association between fungi and plant roots

vascular bundles — small water transporting vessels

water retention capacity — the capacity of soil to keep water

10. INDEX TO SCIENTIFIC NAMES

Acacia
  albida (= *Faidherbia albida*) 7, 8,
  9, 11, 15, 17, 18, 20, 22, 25, 29,
  32, 35, 36, 43, 44, 45, 46, 48, 53
  aneura 37
  arabica (= A. nilotica) 7
  ataxacantha 5, 8, 11, 18, 20, 22,
  29, 51, 54
  caffra 22
  dudgeoni 5, 8, 11, 18, 22, 54
  ehrenbergiana 5, 8, 11, 18, 22,
  37, 45, 54
  elatior 22
  erioloba 22, 37
  gerrardii 9
  giraffae 22
  gourmaensis 5, 8, 11, 18, 22, 55
  hockii 32, 34, 35, 36, 39
  laeta 5, 8, 11, 18, 22, 55
  macrostachya 5, 8, 11, 17, 18,
  20, 22, 29, 51, 55
  macrothyrsa 5, 8, 11, 18, 22, 56
  mearnsii 30
  melanoxylon 9
  mellifera 5, 7, 8, 11, 18, 22, 32,
  34, 35, 45, 56
  nilotica 7, 8, 9, 10, 11, 15, 18,
  20, 21, 22, 29, 32, 33, 35, 36,
  37, 45, 46, 48, 56
  nubica 36
  pennata 5, 8, 11, 18, 22, 57
  polyacantha 5, 8, 11, 18, 22, 58
  senegal 5, 7, 8, 11, 17, 18, 20,
  22, 29, 33, 34, 35, 36, 37, 43,
  45, 46, 47, 58
  seyal 5, 7, 8, 11, 19, 20, 22, 33,
  36, 37, 45, 46, 54, 59
  sieberiana 7, 8, 9, 11, 19, 23, 25,
  29, 32, 36, 39, 47, 60
  spirocarpa (=A. tortilis ssp.
  spirocarpa) 5
  tortilis 5, 6, 7, 8, 11, 13, 14, 17,
  19, 20, 21, 23, 27, 32, 33, 34,
  35, 36, 37, 43, 45, 46, 48, 60
Albizia
  chevalieri 5, 8, 11, 61
  gummifera 35
  lebbeck 5, 8, 11, 19, 43, 46, 61
  lophantha 34
Bauhinia
  rufescens 8, 11, 19, 62
Cassia
  biflora 23

  siamea 5, 8, 11, 31, 36, 43, 46,
  62
  sieberiana 7, 8, 11, 19, 35, 62
Dalbergia
  melanoxylon 5, 8, 11, 63
Dichrostachys
  cinerea 5, 8, 11, 19, 20, 23, 29,
  48, 51, 63
  nutans (= D. cinerea) 7
Entada
  africana 5, 11, 44, 63
Erythrina
  senegalensis 8, 11, 63
Leucaena
  leucocephala 5, 8, 11, 43, 46, 64
Mimosa
  aperata (= M. pigra) 8
  pigra 5, 8, 11, 51, 64
Parkia
  biglobosa 8, 11, 64
Parkinsonia
  aculeata 8, 11, 19, 46, 64
Piliostigma
  reticulatum 8, 9, 11, 19, 20, 65
  thonningii 8, 9, 11, 19, 20, 65
Prosopis
  africana 8, 11, 19, 65
  juliflora 8, 11, 19, 36, 43, 46, 47,
  66
Pterocarpus
  erinaceous 5, 8, 11, 66
  lucens 5, 6, 8, 11, 36, 44, 66
Tamarindus
  indica 7, 8, 11, 19, 20, 29, 67

Reports from the Botanical Institute, University of Aarhus

1. **B. Riemann:** Studies on the Biomass of the Phytoplankton. 1976. 186 p. Out of print.
2. **B. Løjtnant & E. Worsøe:** Foreløbig status over den danske flora. 1977. 341 p.
3. **A. Jensen & C. Helweg Ovesen** (Eds.): Drift og pleje af våde områder i de nordiske lande. 1977. 190 p. Out of print.
4. **B. Øllgaard & H. Balslev:** Report on the 3rd Danish Botanical Expedition to Ecuador. 1979. 141 p. Out of print.
5. **J. Brandbyge & E. Azanza:** Report on the 5th and 7th Danish-Ecuadorean Botanical Expeditions. 1982. 138 p.
6. **J. Jaramillo-A. & F. Coello-H.:** Reporte del Trabajo de Campo, Ecuador 1977—1981. 1982. 94 p.
7. **K. Andreasen, M. Søndergaard & H.-H. Schierup:** En karakteristik af forureningstilstanden i Søbygård Sø — samt en undersøgelse af forskellige restaureringsmetoders anvendelighed til en begrænsning af den interne belastning. 1984. 164 p.
8. **K. Henriksen** (Ed.): 12th Nordic Symposium on Sediments. 1984. 124 p.
9. **L. B. Holm-Nielsen, B. Øllgaard & U. Molau** (Eds.): Scandinavian Botanical Research in Ecuador. 1984. 83 p.
10. **K. Larsen & P. J. Maudsley** (Eds.): Proceedings. First International Conference. European-Mediterranean Division of the international Association of Botanic Gardens. Nancy 1984. 1985. 90 p.
11. **E. Bravo-Velasquez & H. Balslev:** Dinamica y adaptaciones de las plantas vasculares de dos cienagas tropicales en Ecuador. 1985. 50 p.
12. **P. Mena & H. Balslev:** Comparación entre la Vegetatión de los Páramos y el Cinturón Afroalpino. 1986. 54 p.
13. **J. Brandbyge & L. B. Holm-Nielsen:** Reforestation of the High Andes with Local Species. 1986. 106 p.
14. **P. Frost-Olsen & L. B. Holm-Nielsen:** A Brief Introduction to the AAU - Flora of Ecuador Information System. 1986. 39 p.
15. **B. Øllgaard & U. Molau** (Eds.): Current Scandinavian Botanical Research in Ecuador. 1986. 86 p.
16. **J. E. Lawesson, H. Adsersen & P. Bentley:** An Updated and Annotated Check List of the Vascular Plants of the Galapagos Islands. 1987. 74 p.
17. **K. Larsen:** Botany in Aarhus 1963 - 1988. 1988. 92 p.

AAU REPORTS:

18. Tropical Forests: Botanical Dynamics, Speciation, and Diversity. Abstracts of the AAU 25th Anniversary Symposium. Edited by **F. Skov & A. Barfod.** 1988. 46 pp.
19. Sahel Workshop 1989. University of Aarhus. Edited by **K. Tybirk, J. E. Lawesson & I. Nielsen.** 1989.
20. Sinopsis de las Palmeras de Bolivia. By **H. Balslev & M. Moraes.** 1989. 107 pp.
21. Nordiske Brombær (Rubus sect. Rubus, sect. Corylifolii og sect. sect. Caesii). By **A. Pedersen & J. C. Schou.** 1989. 216 pp.
22. Estudios Botánicos en la "Reserva ENDESA" Pichincha - Ecuador. Editado por **P. M. Jørgensen & C. Ulloa U.** 1989. 138 pp.
23. Ecuadorean Palms for Agroforestry. By **H. Borgtoft Pedersen & H. Balslev.** 1990. 120 pp
24. Flowering Plants of Amazonian Ecuador - a checklist. By **S. S. Renner, H. Balslev & L. B. Holm-Nielsen,** 1990. 220 pp.
25. Nordic Botanical Research in Andes and Western Amazonia. Edited by **S. Lægaard & F. Borchsenius,** 1990. 88 pp.
26. HyperTaxonomy - a computer tool for revisional work. By **F. Skov,** 1990. 75 pp.
27. Regeneration of Woody Legumes in Sahel. By **K. Tybirk,** 1991. 81 pp.